Table of Contents

Chapter		Page
I	Sources of Water and Their Characteristics	I-1
II	Water Production, Storage and Distribution	II-1
III	Public Health	III-1
IV	Operational Mathematics	IV-1
V	Applied Hydraulics	V-1
VI	Reservoirs	VI-1
VII	Pipe, Types, Uses and Protection	VII-1
VIII	Pipe Installation and Maintenance	VIII-1
IX	Pipeline Appurtenances	IX-1
X	Meters and Services	X-1
XI	Pumps and Motors	XI-1
XII	Electricity, Instrumentation and Control	XII-1
XIII	Safety	XIII-1
XIV	Chlorine Handling and Safety	XIV-1
XV	Maps, Drawings and Records	XV-1
XVI	Public Relations	XVI-1
XVII	System Operation	XVII-1
	Glossary	G-1
	Bibliography	B-1

Introduction

A reliable water supply is one of the most important elements of a community. Not only must sufficient quantities of water at adequate pressure be available but the water must also be safe for human consumption. A water distribution operator is legally and morally responsible to see that this is accomplished. He must remember the importance of water to a community and produce it in the most efficient, reliable, and economical manner.

A water distribution operator deals with the water system daily and is responsible for giving the best service possible at all times. The operator is the first line of contact between the general public and the company for which he works. Therefore, he must not only know and apply critical water-distribution techniques, but he must also conduct himself in such a manner that the general public is entirely confident that a vital service is functioning properly.

This handbook is intended to be an introduction to water distribution. As one reads it, he will undoubtedly discover the reasons for many things that are accomplished routinely every day. In some places the text has more than is needed and yet on other subjects, it may be shy of knowledge for specific needs. The operator is encouraged to study other reference material in those areas, some of which are listed in the bibliography. As one reads this book, he should think always of his own system, trying to understand not only the parts of the system but the system as a whole.

Almost everyone, including some persons in the water industry, takes for granted the supply of water. This is largely because of the results of the water-distribution operator, and it will continue to be so as long as he does his work in a professional manner.

In the past, water utilities have striven to provide a safe, abundant water supply with adequate pressure and at a minimum cost. The industry has been successful in achieving these aims, and water is one of the least expensive vital resources available. These requirements will continue to be met in the future as new ones are constantly being added.

Safeguarding the quality of water will become increasingly important. A water-distribution operator's daily activities can greatly influence the quality of the water that is delivered. Therefore, an operator is a vital person in the campaign to provide *Better Water for People*.

Chapter I
Sources of Water and Their Characteristics

Introduction

This chapter is designed to give the distribution operator an understanding of the sources of water and the characteristics of those sources. Although roughly three fourths of the earth's surface is water, only a small portion is normally used for public water supply. On a volume basis, the salt water in the oceans composes over 97 per cent of the earth's water. Of the remaining 3 per cent which is fresh water, 75 per cent is more or less permanently stored in the ice caps and about 25 per cent is stored beneath the surface of the land. However, only a small portion of this groundwater can be withdrawn in significant amounts. Only one third of 1 per cent of fresh water is in lakes and rivers and one fifth of this amount is located in the Great Lakes.

Hydrologic Cycle

Unlike other commodities, water is not manufactured. Instead, it is continuously recycled naturally. This cycle, called the hydrologic cycle, involves evaporation, precipitation, transpiration, infiltration, percolation and runoff.

An understanding of the hydrologic cycle helps explain the way in which all water becomes available. Water vapor continuously accumulates in the atmosphere as water evaporates from oceans, lakes, ponds, reservoirs, streams, land, and falling rain. Plants also release moisture into the atmosphere in a process called transpiration. The combined effect of evaporation and transpiration is called evapotranspiration. When a mass of warmer air carrying sufficient water vapor is forced upward into cooler levels, the vapor condenses and forms clouds, which eventually release the water as precipitation. Although rain is the most common form of precipitation, snow, hail, and sleet are others. Precipitation that falls on land surfaces may be absorbed by the soil (infiltration) and thus recharge the groundwater supply, or it may run off and form rivers and streams, which flow into oceans and lakes. The absorbed water also moves through the ground by the process of percolation and may either remain in the ground or eventually seep into surface waters such as those just mentioned. Little or no new water is produced in nature, but rather water is continually cycled through many different forms. See Fig. I-1.

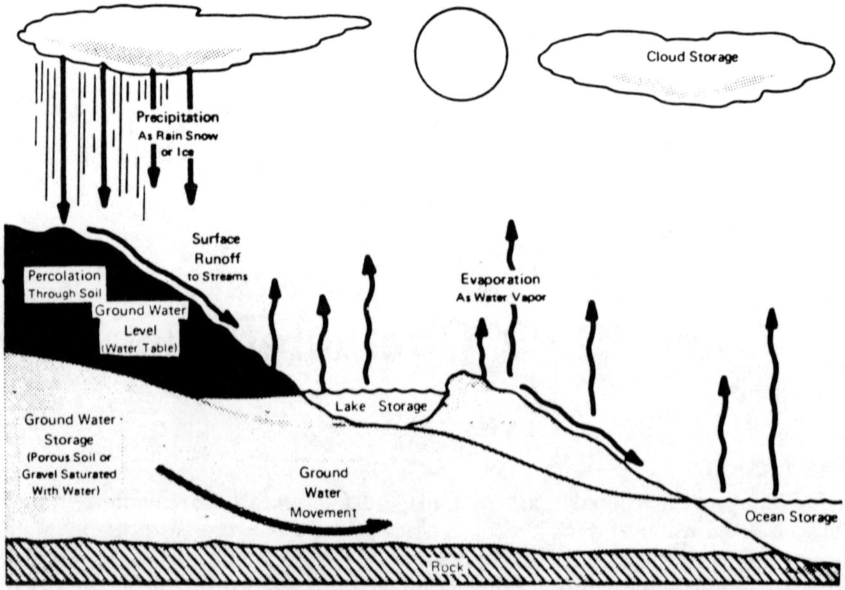

Fig. I-1. Hydrologic Cycle

Surface Water

Although only one quarter of public water systems rely on surface water, those systems serve three quarters of the populace. Surface water sources are normally rivers, lakes, or impoundments such as man-made reservoirs.

River water is normally contaminated by pathogenic bacteria and generally requires extensive treatment before being suitable for distribution. The belief that streams are self-purifying within varying distances refers primarily to the destruction of organic wastes by oxidation and is a myth with regard to eliminating disease. River water is seldom uniform in quality and may vary from day to day. Furthermore, rivers may not provide enough water when the need is greatest during the summer months. Because of this, storage facilities such as open or covered reservoirs, tanks, or man-made lakes may be necessary. These impoundments can be used to store water for use during dry periods, to average out peak or low flows, and to provide water to locations where the supply is inadequate.

Large lakes may yield water of fine quality except near shore lines and sewer outlets. Depending on conditions, the storage time of water in a lake may allow natural processes largely to reduce levels of biological and bacteriological contamination. However, several factors may negatively influence the quality of lake water. Winds may blow directly from the inlet to the outlet, establishing streams in the lake which carry the inlet water directly to the outlet, thereby eliminating the benefits of longer storage. Water quality may suffer because of a seasonal overturn of water, which causes the bottom water to exchange places with the top water and stirs up sediment in the process. Also, algae growth may impart an objectionable taste or odor to the water. However, if these conditions and others do not occur, or are effectively controlled, lake water may be uniform in quality from day to day and require a minimum of treatment.

All surface supplies of water are exposed to accidental, intentional or incidental pollution by humans and without some form of treatment should be presumed to be unsafe for human consumption. Also, whether a river, lake, or man-made impoundment, any surface water source must be evaluated for its safe yield. Safe yield is the maximum rate at which water can be withdrawn continually during a period of years in which the probable driest period is likely to occur. Safe yield is a function of storage provided and generally ranges from lowest dry-weather flow without storage to roughly mean annual flow with heavy storage development.

Ground Water

Ground water is defined as that water contained in the saturated subsurface zone lying immediately below the water table. Two categories of ground water are of special interest in water supply: free ground water and confined ground water. Free ground water is located below the water table and is normally underlain by an impervious stratum. It moves obliquely downward according to the slope of the water table. Confined ground water is ground water that is confined above and below by impervious strata. If it is under positive pressure at a given location, it is the source of artesian water if tapped. See Fig. I-2.

Spring water occurs where a water-bearing stratum reaches the ground surface. An impervious layer above a water bearing stratum may also serve as geological protection for ground water.

Ground water is often clear and free of organic matter because of the filtration effect of water percolating through soil, sand, or gravel. Its quality and mineral content are normally constant over long periods, and, because of natural filtration, it may be relatively free from bacteria. Color is infrequently found in ground water and is cause for suspicion when it is present.

The mineral content of water obtained from shallow wells will vary depending on the mineral characteristics of the soils in the area. Water from deep wells has certain advantages over that from shallow wells because it is usually clearer, contains fewer bacteria, is more uniform in mineral content, and is usually less susceptible to contamination. Water from deep wells almost always contains concentrations of dissolved minerals but is normally clear and colorless because of filtration. Waters from the same well field are generally of similar quality and temperature, but a disadvantage is their frequently high mineral content.

Spring water is usually similar in quality to the water found in nearby wells, but the flow of water to a spring may be rapid and natural purification may not occur. Thus, spring water should always be regarded with suspicion.

Other Water Sources

Seawater desalination and wastewater reclamation are other sources for potable water. Desalination involves the conversion techniques of distillation, ion exchange, freezing, and a process using permeable membranes called reverse osmosis. Because of its high cost, desalination is not presently economically attractive for must municipal water supplies. As reclamation technology is improved and wastewater requirements become more stringent, the outlook for

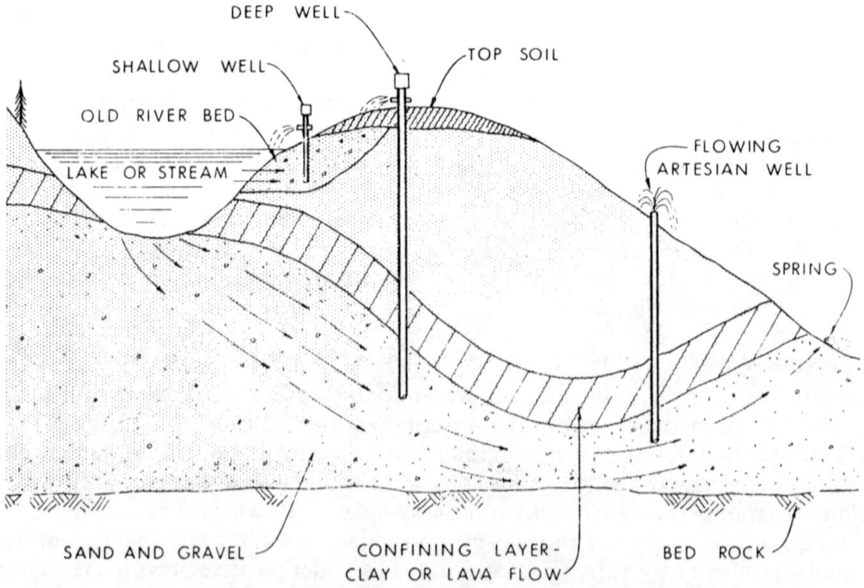

Fig. I-2. Geologic Section Showing Occurrence of Ground Water

wastewater reclamation should become more favorable. At present, reclaimed water is used primarily for agricultural, industrial and recreational purposes, and the use of reclaimed water for domestic purposes is generally not practical. Public acceptance and certain technical achievements must be obtained before proposals for the use of reclaimed water can be pursued.

Water Quality Variables

There are several water quality variables with which a water utility employee should be concerned. Some of these quality variables involve hygienic considerations and others, aesthetic considerations. However, all these variables are important if water is to be delivered in a safe, clear, good tasting, and pleasing condition to the customer. Some of the primary quality variables are pathogenic organisms, turbidity, total dissolved solids, tastes and odors, color, hardness and pH.

Pathogenic organisms that can be transmitted readily by water are those that cause typhoid, dysentery, cholera, and gastroenteritis. These will be covered more completely in Chapter III.

Turbidity refers to suspended particles in the water and may be caused by micro-organisms and nonliving organic or inorganic matter. Water that contains certain chemicals such as manganese, iron, or sulfur may cloud upon exposure to air after being allowed to remain idle. Total dissolved solids (TDS) differs from turbidity in that it is a measure of nonsettleable and nonfilterable solid particles in solution.

Tastes and odors are normally caused by minerals, dissolved gases, earthy substances (especially from a newly constructed earth reservoir), decayed vegetation, industrial wastes, and plankton such as algae and higher forms of bacteria.

Color in water may be of animal, mineral, or vegetable origin and may be attributed to metallic substances (such as iron compounds), humus material, peat, algae, weeds, or protozoa. Water should not be judged by appearance alone as highly nitrated or polluted waters are often clear, bright, and free from color.

Hardness of water is a characteristic that received its name from the consumer when people noticed that in certain waters it was difficult to make soap lather. Hardness is caused by the presence of calcium and magnesium salts in solution, these minerals having dissolved from deposits through which the water has percolated or drained. Besides causing lather problems, hard water is also responsible for grey clothes and the ring that can form around bath tubs and sinks. Moreover, it may be unsuitable for some manufacturing processes by causing incrustation in steam boilers and piping systems and increasing corrosion.

The letters pH constitute a symbol used to designate the relative acidity or alkalinity of water. The pH scale extends from 0 to 14 with 7 as neutral. It is a logarithmic scale, so each whole number represents a ten-fold increase over the previous one. Numbers below 7 indicate acidity and numbers above 7 indicate alkalinity. The further from 7 in either direction, the more severe the condition.

Chapter II
Water Production, Storage and Distribution

Introduction

This chapter is intended to acquaint the reader with surface- and ground water. In particular, this chapter discusses water production and supply, transmission and distribution, and the utilization of water.

Imported Water

Previously, it was noted that the natural water supply for any region is primarily dependent upon either surface water or ground water. However, in semiarid and highly populated areas, local water sources are usually insufficient for the needs of the population. The alleviation of this problem has been accomplished by importing water by way of some of the world's greatest engineering feats dating back to the aqueducts of Rome. Presently, water for southern California is imported over 300 mi from the Colorado River and 500 mi from rivers in northern California. Many large areas of population throughout the country require imported water.

Ground Water Production

Wells are normally classified in a variety of ways. One classification pertains to depth. Previously, wells less than 100 ft deep were classified as shallow wells and wells deeper than 100 ft were classified as deep wells. However, wells are presently considered deep wells if they penetrate a geological formation of impervious material. Wells that do not penetrate impervious strata are considered shallow wells.

Wells are also classified by their method of construction. They may be dug, bored, drilled, or driven. Dug wells are usually shallow and are formed by the use of hand tools or a "clam shell" excavator. These wells are normally large in diameter (up to and greater than 30 ft) and are relatively vulnerable to contamination. Bored wells are formed by the use of an auger and are seldom used any more. Drilled wells are formed by either percussion or rotary drills. Drilling is the method most commonly used to produce a well. Deep drilled wells normally provide excellent protection against contamination if properly constructed. Driven wells consist of a series of pipe lengths fitted with a well point on one end. The well point is driven through the subsurface by heavy hammer blows. Water enters through a screened opening in the shank. If the underground is free of rocks and the water table is not more than 50 or 60 ft deep, then this is a simple and inexpensive method. Of the four types, only the drilled well dependably provides water in large amounts.

When water is pumped from a water-bearing aquifer, it has the effect of draining a portion of the aquifer. This portion is an inverted cone shaped area called the *cone of depression*. The peak of the cone is at the pumping level—the depth to which the water recedes after the well has been pumped long enough for the water to stabilize at one elevation. See Fig. II-1.

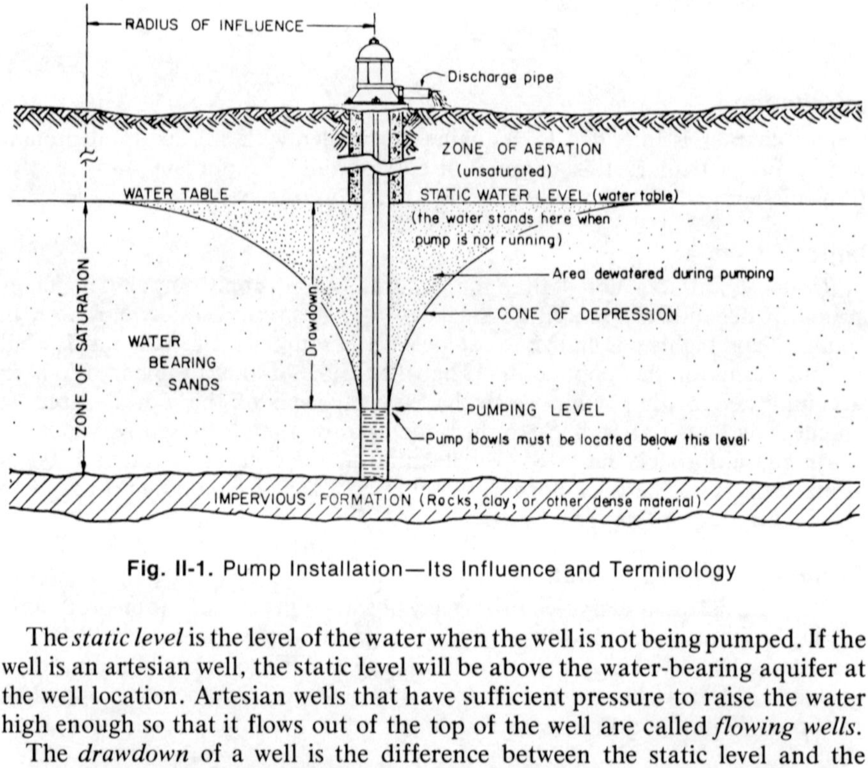

Fig. II-1. Pump Installation—Its Influence and Terminology

The *static level* is the level of the water when the well is not being pumped. If the well is an artesian well, the static level will be above the water-bearing aquifer at the well location. Artesian wells that have sufficient pressure to raise the water high enough so that it flows out of the top of the well are called *flowing wells*.

The *drawdown* of a well is the difference between the static level and the pumping level. The *radius of influence* is the area from the well to the edge of the top end of the cone of depression. If two wells are close enough so that their circles of influence overlap, then *mutual interference* occurs and the production of both wells is decreased. See Fig. II-2.

Measurements. The measurement of static and pumping water levels can be accomplished electrically, mechanically, or pneumatically. These measurements enable the water operator to diagnose the condition of the well. If the static level drops but the drawdown remains the same, then the operator should know that the water table is falling. This means that the aquifer is being depleted faster than it can recharge itself. If the static level is unchanged but the drawdown has decreased, then there may have been a loss of efficiency of the pump. If the static level is unchanged but the drawdown has increased, then the screen or strainer is

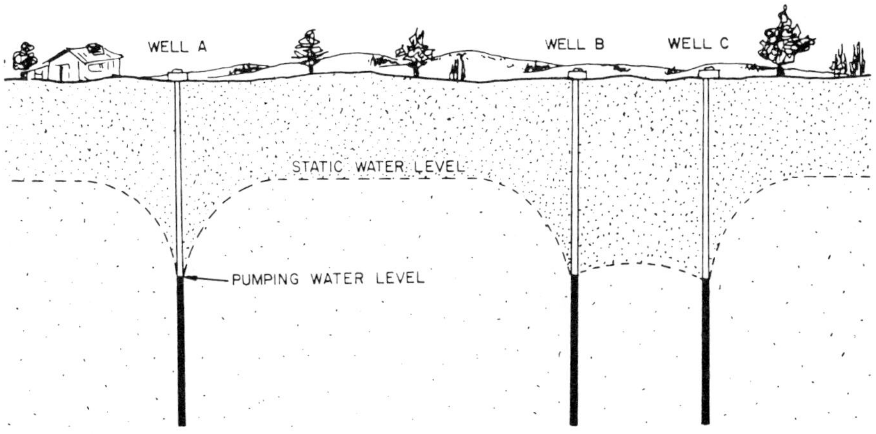

Fig. II-2. Mutual Inference Existing Between Wells B and C

clogged and water is not freely flowing into the well. Inspection by television cameras may show exactly where the trouble lies. The well should be surged or flushed out. If this fails it must be acid-cleaned or swabbed. When measuring the static or pumping levels the operator should allow sufficient time for the true levels to be reached. A careful record should be kept on each well recording static and pumping levels and the rate of production. Water meters are one of the best and most accurate ways to record production, but production may also be closely approximated by multiplying the time of operation by the output of the pump (see Chapter IV).

Yield. The yield of a well is dependent upon several factors, such as the depth into the aquifer, the composition of the aquifer, the diameter of the well, the temperature of the water, and the amount of water flowing through the circle of influence. Increasing the diameter of a well is seldom worthwhile because, normally, doubling a well diameter only increases production by 10 to 15 per cent. The yield of deep wells is normally independent of local conditions as the area of recharge is usually some distance away. These aquifers serve much like storage areas by averaging out the fluctuations in rainfall or snowfall. For this reason the yields of deep wells are normally constant. Shallow wells are not normally constant because they depend on local rainfall conditions for recharge.

All ground-water supplies must be adequately protected from contamination or pollution. It is the responsibility of the water utility employee to be on the alert constantly for conditions that can lead to possible contamination.

Note: All ground-water withdrawal points should be located a "safe" distance from sources of pollution. Sources of pollution include septic tanks, and other individual or semipublic sewage disposal facilities, sewers, farm animals, fertilizers, and pesticides. Where water resources are severely limited, ground-water aquifers subject to contamination may be used for water supply if adequate treatment is provided.

Sanitation Requirements. It is difficult to set fixed safe distances, but, generally, a safe distance is such that under the most adverse conditions no contamination or pollution should occur.

Typical suggested distances:

	Distance From Community Water Supply Wells—*ft*
Sewer, watertight septic tank, or pit privy	50
Subsurface sewage leaching field	100
Cesspools or seepage	150

Where, in the opinion of the enforcing agency, adverse conditions exist, the above distances shall be increased or special means of protection, particularly in the construction of the well, shall be provided.

The casinghead (a watertight seal at the top of the well) and well vents should be properly constructed and maintained to prevent contamination. For safety, and as a precaution against possible contamination of the aquifer region, unused wells should be capped and abandoned wells should be filled. Pump houses should be kept clean, adequately drained, and protected against flooding.

Pros and cons. Generally, ground waters are clearer (but not necessarily purer), cooler, and have a more constant temperature than surface waters during the summer. In winter, ground waters are not nearly as likely to freeze as are surface waters. The dissolved minerals in water from the same well are usually uniform. Two important advantages of ground water are that storage reservoirs and large water transmission lines are not needed.

Ground water sources are therefore suitable for small communities, private estates, and isolated institutions and industrial plants. However, there are some disadvantages to using ground water sources.

In general, ground waters contain higher concentrations of dissolved minerals than surface waters in the same area and are usually "harder" than surface waters.

Sometimes, well water will contain hydrogen sulfide, which makes the water smell like rotten eggs. Also, there may be *Crenothrix* bacteria, which thrive on water containing iron, or the well may contain corrosive water that oxidizes steel and cast-iron pipes. The result is a red rust in the water or "red water." Many well waters contain iron and manganese, which may discolor the water.

Pumping well water is usually more expensive than pumping surface water. And, if the water supply comes from several wells, each of which produces water of different chemical characteristics, then the characteristics of the blended water are changed if wells are added or taken out of service.

Surface Water and Impoundments

Surface water and impoundments present different problems for a water utility. Surface-water supplies are always susceptible to contamination or pollution by cities, industries, agriculture, birds, and animals. Therfore, they must always receive treatment before domestic use. Treatment with chemicals such as copper sulfate (bluestone) may be used to prevent the growth of algae that can cause unpleasant tastes and odors.

The watershed area of any surface supply should be inspected regularly for sources of possible contamination or pollution. Erosion on the watershed should be prevented or limited to reduce the turbidity of the water and the silting rate.

Unauthorized recreational use of reservoirs and lakes is a problem in some areas. All recreational use of water-supply lakes or reservoirs should be controlled to prevent harmful effects on water quality. If possible, reservoirs should be fenced to prevent unauthorized use. On combination recreation-water-supply reservoirs, a barrier should be erected across the reservoir to prevent possible contamination near the outlet. Because recreational waterways in many parts of the world and North American are few, the water-utility employee should understand and be ready to explain restrictions. An informed public is usually an understanding public.

Transmission, Distribution, and Operational Storage

In all but the largest systems, the greatest determining factor of the amount of storage and the size of distribution mains is the fire-flow demand, which is usually the flow necessary to combat a major fire. This is especially true in small communities. The water normally used for firefighting purposes is a small amount compared to the annual consumption of water. However, it is necessary for a system to be of adequate size and capacity when a fire occurs if a community is to receive fire insurance at a low rate and to fight fires effectively.

Water-distribution systems are normally designed in grid or looping designs to avoid lines that dead-end. System hydraulics and reliability are also improved by a loop line. A dead-end line must be flushed frequently or the customers on that line will be served a poor quality water. The waste of water caused by flushing is avoided with a grid system. These systems also allow for greater water flow into an area during times of high demand or firefighting needs. See Fig. II-3.

Another type of distribution system is the tree system, which consists of a single arterial main which is decreased in size as flow demands decrease. Branches depart at right angles, with sub-branches from each branch.

All distribution systems should contain sufficient valving so that areas may be isolated. This would allow for the minimum disruption of water service during times of leaks or maintenance. The valves also allow for certain areas to be isolated if contaminated. It is important that all valves be operated periodically to ensure their performance.

The minimum pressure desirable in a system is normally 20 psi. Many modern water-using appliances or machinery will not function properly on pressures lower than 20 psi. Often a water agency will have two or more service levels, or pressure zones, in areas of varying topography to avoid excessive pressure in certain areas. This might be accomplished by pumping the high zone or reducing pressure to the low zone. Continuous positive pressure is necessary in all distribution systems to prevent creating a vacuum in the water-supply main. These vacuums can lead to back siphonage with possible contamination or even collapse of the pipe sections. Backflow prevention will be discussed in later chapters.

Operational storage is necessary to average out high and low demands exerted on the water source. This storage prevents the on-and-off operation (cycling) of auxiliary pumps during peak demands. Storage may be elevated or it may be at

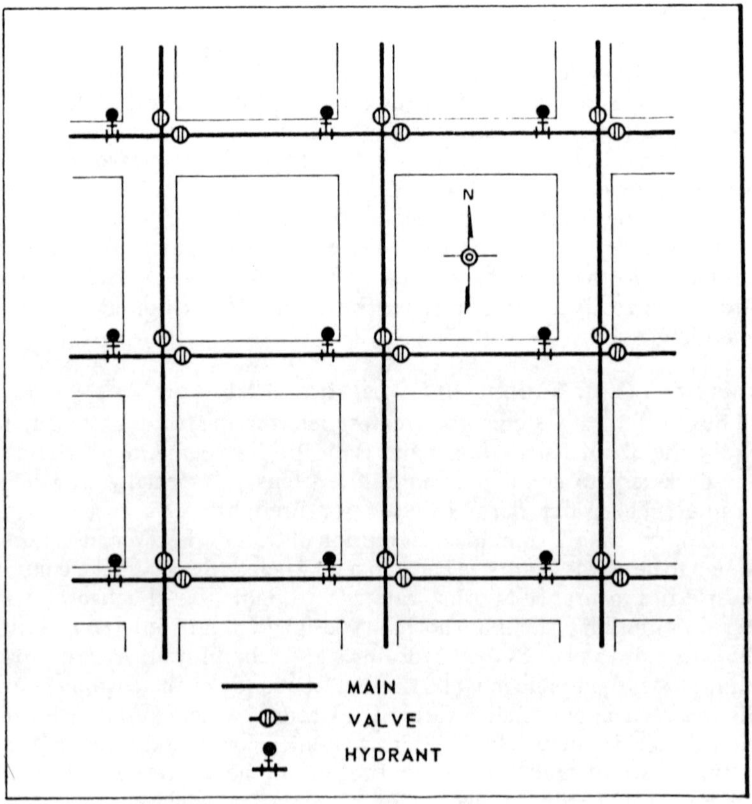

Fig. II-3. Typical Distribution Grid Layout

ground level; however, elevated storage is preferred because it gives the required pressure for a system regardless of whether the pumps are operating or not. This is especially important during fires. Figure II-4 shows how distribution reservoirs function with respect to demand.

The most common ground storage is a clear well (treated water storage). These are located at treatment plants and serve to reduce fluctuation in treatment flow rates. The uniform operation of treatment plants and low-lift pumps yields greater efficiency and lower power costs.

Maintenance of storage reservoirs is necessary to protect against corrosion or other deterioration and contamination. All signs of leaks should be investigated. The reservoir should be drained occasionally, cleaned, and painted if necessary. The surrounding areas of reservoirs should be shaped to avoid any possible contamination. Surface and underground drainage should be in a direction away from the reservoir.

Piping, meters, valves, and pumps will be discussed more thoroughly in later chapters.

Utilization of Water

There are several factors that can lead to increased consumption in any water

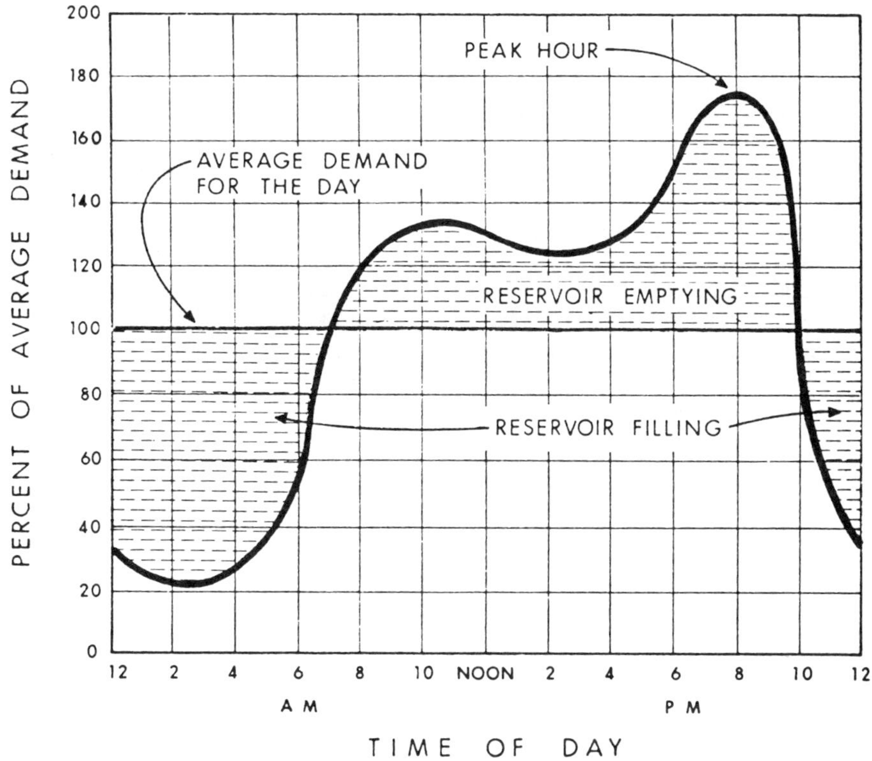

Fig. II-4. Daily Variation of System Demand

system. Some of these are good quality water, low rates, good service, high pressure, climate, and the absence of meters. During a water shortage, reducing water pressure is often the simplest and easiest method to decrease consumption. Normally, 5 to 10 per cent of the potable water supplied is lost due to leakage, the remaining water usage consists of 45 per cent for domestic use, 45 per cent for commercial use, and 10 per cent for public use.

Typical Variation in Water Consumption Rates
(as percentage of annual average rate)

	Population	
Rate	Less than 100 000	Over 100 000
Annual Avg.	100	100
Monthly max.	120-150	110-130
Daily max.	150-250	125-175
Hourly max.	300-400	200-300
Monthly min.	75-90	80-80
Daily min.	50-75	60-80
Hourly min.	25-50	50-75

Recreational Use of Domestic Water-Supply Reservoirs

Water utilities in the US and Canada provide water of the highest quality in the world. To improve upon this, each water utility must continue to recognize its responsiblity to deliver a safe and appealing product to its customers. A growing demand for the use of reservoirs for recreational purposes, however, may make this responsibility more difficult to fulfill in the future. In some areas, uncontrolled use of reservoirs has resulted in deterioration of water quality. Without strict controls governing recreation, waterborne diseases could become a serious problem.

It should be the vested responsibility of water utilities to determine the use of their reservoirs. The decision concerning recreation should be based on (1) state (provincial) and local laws and regulations, (2) information provided by AWWA and other organizations that conduct research and present technical reports, and (3) an assessment of the public's needs for the full utilization of water resources.

Reservoirs used for recreation must receive close supervision. In all cases, water from reservoirs used for recreation must be treated. Recreational use of reservoirs will result in added capital and operating costs that may increase the charges for water service.

Public water-supply reservoirs should not be used for recreation if other surface waters are available. Body contact sports such as swimming should not be allowed in public water-supply reservoirs. Distribution reservoirs, from which water is supplied directly to the public, require the strictest control and under no condition should be used for recreation.

Chapter III
Public Health

Introduction

This chapter is designed to introduce the reader to some of the factors that affect water quality and are therefore important to those concerned with the public health. It is the responsibility of the water utility personnel to deliver a safe water to the consumer. The distribution operator plays a key role in this service.

Contamination and Pollution

Water is considered as the universal solvent because almost everything will dissolve to a limited extent in water. This presents problems for delivering a safe water to the consumer. A "safe" water is water that can be consumed without menace to one's health. A "potable" water is water that is satisfactory for drinking purposes with regard to the water's chemical, physical, and biological characteristics.

Water in nature is never pure. It always contains mineral matter in solution or suspension and dissolved gases. Water may be polluted or contaminated. Pollution is the introduction of substances into the water in sufficient quantity to affect the original quality of the water, to make it objectionable to sight, taste or smell, or to make it less useful. Contamination is the introduction of toxic materials, bacteria, or other harmful agents into the water that make it hazardous and therefore, unfit for human consumption.

Although rainwater may approach purity in the latter stage of a storm, almost all rainwater contains dissolved gases, bacteria, and some radioactivity. As water flows on the surface or in the ground, it collects mineral and humus material from the soil. The water also contacts pollutants from domestic and industrial wastes.

Health Standards

The US Environmental Protection Agency (EPA) provides minimum standards for water quality. These standards set limits on the physical, chemical, biological, and radiological nature of water. Physical characteristics are normally based on tests that provide units of measure. Chemical characteristics are measured and reported on the basis of weight per unit volume of water; that is, as so many parts per million (ppm) or milligrams per liter (mg/l). The terms ppm and mg/l will be explained more completely in Chapter IV. At this point the operator need only remember that 1 ppm of a chemical in water means that for every 1 million gal (8 340 000 lbs) of water, 8.34 lb of the chemical are present. In water, 1 mg/l is equivalent to 1 ppm.

The water utility is held responsible for the water quality from the source of water supply to the customer's service connection. It is therefore necessary for the utility to maintain the water quality throughout the system and to prevent the existence of health hazards.

Water Quality Factors

The source of a water supply must be protected against contamination or pollution. As part of this safeguarding the EPA requires that a systematic program for conducting sanitary surveys be implemented. A sanitary survey is normally made by a trained technician and includes physical and chemical tests on the water quality. In the watershed of surface supplies the survey should check for stagnant ponds that could be sources of algae or bacteria entering the main supply. Also, the survey should ensure that sewage does not flow into the water supply because of defective sewer or septic systems. For well waters, the sanitary survey should check for abandoned wells and borings, leaky well casings, poor design or location, and other sources of possible contamination. Especially important is any source of contamination within the zone of influence. Flooding also may cause contamination of a water supply. (See Fig. III-1 for method used for proper sampling.)

Aesthetic aspects of water such as taste, odors, color, and temperature are actual safety of the water, they may cause people to choose an alternative source of water that could be unsafe. In an actual instance, a town was supplying water that was safe but smelled bad and was frequently colored. The people in the town began drawing water from a well in the local park. The water was clear and odor free, but was also highly contaminated.

Water Contaminants

Contamination of water may be biological, chemical, or radiological. Biological contamination may be by bacteria or viruses. Bacterial contamination is of special concern to the water utility because pathogenic bacteria that are transmitted into the water may cause many different types of disease. The most common waterborne sicknesses are typhoid, cholera, diarrhea, and dysentery.

Although bacteria are the most plentiful of all living organisms, only a few are pathogenic. Many bacteria are helpful and even necessary for everyday living. The problem is to identify the presence of harmful bacteria in the water. This is done by the use of indicator organisms.

Disease-causing bacteria reside in the intestinal tract of humans and animals. The disease bacteria are almost impossible to test because they represent only a small proportion of the total bacteriological count. The greatest proportion of intestinal bacteria is the coliform group. Tests for the presence of coliform bacteria are relatively simple. The presence of coliform bacteria and specifically E. Coli (a particular type of coliform bacteria) is an indication that fecal matter is present, even if in small quantities. This is cause for concern because pathogenic bacteria may also be present.

Representative water samples are taken regularly throughout the entire **system. Field** personnel assist the laboratory by taking samples as required.

Collecting Water Samples for Bacteriological Analysis

In order to avoid accidental contamination of the sample, the procedure for taking samples in the sterile plastic (polypropylene) bottles is herein depicted for strict compliance by all personnel:

Flame the mouth of sample tap of faucet with Butane torch.

Allow water to run at least five minutes to flush line; longer, if necessary to draw a fresh sample through the delivery line.

After line has been flushed, determine water temperature and record on sample sticker.

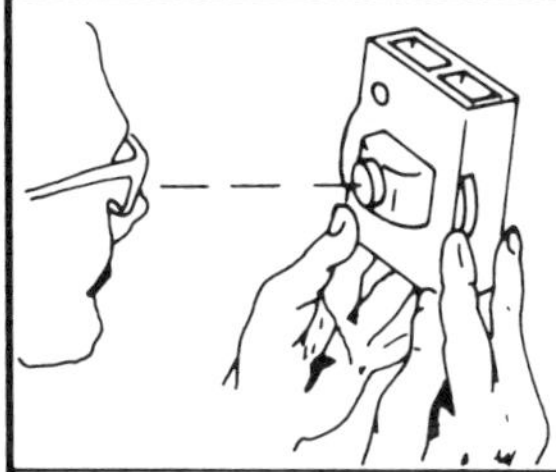

Determine residual chlorine. This data will indicate when a fresh representative sample is flowing through the line.

Remove cap from sample bottle. Avoid touching lip of bottle or inside of cap when removing or replacing. Fill bottle to black mark. Never fill bottle completely.

Place cap on bottle with care to avoid contaminating the water and **screw cap on tightly.** Place identification tag on each bottle, showing location, date, time, temperature, residual chlorine, initials of sampler, etc. Keep all sample bottles in a vertical position while transporting them to the laboratory. Samples should be stored at a temperature of 40° - 50° F. and delivered to laboratory **as soon as possible, preferably the same day.**

NOTE:

The screw caps on the plastic bottles are not screwed down tightly when the bottles are sterilized. If the cap should become separated from the bottles in the field, please return empty bottles and cap to laboratory for re-sterilization. Plastic bottles marked "PL" have not been sterilized, and they are to be used only for taking plankton samples.

Fig. III-1. Typical Sampling Instructions

So that bacteria in sample will not be destroyed, a dechlorinating agent is placed in the sample bottle. For this reason the sample bottle must not be rinsed out before taking the sample.

One of the tests for E. Coli is the multiple fermentation tube method, which is standard for all types of waters. The principle is that bacteria will ferment the sugar lactose in a nutrient liquid and produce a gas within 48 hr. The presence of coliform bacteria is confirmed by a second test with brilliant green bile broth, which inhibits the growth of any bacteria other than the coliform group. The degree of contamination is determined by a mathematical formula that yields the most probably number of "MPN," an estimate of the number of bacteria.

A second test method is the membrane filter method. The membranes are very thin films of cellulose that stop the bacteria. After 18-22 hr at 35C (121F), colonies of bacteria develop and are counted by stereomicroscope. This is a simpler and quicker method of detection, and, with special equipment, it may be performed in the field.

Tests and procedures for the presence of coliform bacteria are explained and outlined in *Standard Methods for the Analysis of Water and Wastewater. Standard Methods* is published jointly by AWWA, the American Public Health Association, and the Water Pollution Control Federation and is considered the authoritative manual on the subject of water testing. Another reference is *Simplified Procedures for Water Examination M-12,* which is also published by AWWA.

Viruses of particular concern to a water utility are those that cause infectious hepatitis and polio and those causing gastrointestinal disease.

Several chemical aspects of water are significant. Calcium and magnesium cause hardness in water. Alkalinity is important for corrosion control. Iron and manganese may cause staining of clothes and of plumbing fixtures. Chlorides cause the water to taste salty. Excessive amounts of flouride may cause staining (mottling) of children's teeth (however, in lower concentrations of about 1 ppm it may help to prevent cavities). Nitrates and nitrites are an indication of chemical fertilizer pollution and can be very harmful to infants by causing a condition known as "blue babies."

Regulatory agencies having jurisdiction in an area always should be consulted for prescribed water quality characteristics. However, according to EPA regulations, waters should comply with the following criteria.

TABLE III-1

Criteria Established by EPA for Maximum Concentrations of Listed Substances

Substance	Concentration—*ppm* (*mg/l*)
Foaming agent (MBAS)	0.5
Chloride (Cl)	250.0
Copper (Cu)	1.0
Carbon chloroform extract (CCE-m)	0.7
Iron (Fe)	0.3
Manganese (Mn)	.05
Nitrate-N + Nitrite-N	10.0
Sulfate (SO4)	250.0
Zinc (Zn)	5.0
Total Dissolved solids	500.0

TABLE III-2

Minimum Criteria for Rejecting a Water Supply

Substance	Concentration—*ppm* (*mg/l*)
Arsenic (As)	0.05
Barium (Ba)	1.0
Cadmium	0.01
Chromium (hexavalent) (Cr_6)	0.05
Lead (Pb)	0.05
Mercury (Hg)	0.002
Nitrate (as N)	10.0
Selenium (Se)	0.01
Silver (Ag)	0.05

Radiological contamination may be caused by radioactive materials used in research and medicine, production of radioactive material, and by radioactive wastes and fallout from atomic tests. As yet, not enough radioactivity has been found in water to cause a health hazard.

Water Treatment

The hydrological cycle and the effect it has in supplying water and the storage effect were discussed in Chapter 1. These are examples of natural means of improving water quality. Flowing streams also have some effect, but not nearly the effect that some people seem to believe. Flowing streams usually have more effect on organic matter and little effect on bacteria. Important factors affecting natural purification are time, temperature, sunlight, velocity of flow, reaeration, and amount of dilution with clean water.

Artificial treatment normally consists of several steps. Each step is performed in a certain sequence to minimize the cost. Although the distribution operator is seldom involved directly with water treatment, he should have a basic understanding of the process.

Chemicals are added to water entering a treatment plant. Some of the chemicals are coagulants such as alum, lime, ferric chloride, ferric sulphate, and polyeletrolytes. Their purpose is to help remove suspended matter by causing small particles to clump together into heavier or larger pieces. The addition of these chemicals is called the coagulation phase.

Following the coagulation phase, water flows into a basin where the water and chemicals are thoroughly mixed. The purpose of this is to mix the coagulants with the water and to create clumps of heavier material called "floc." This is the flocculation phase.

After coagulation and flocculation, the water flows into a sedimentation basin and remains from 1 to 4 hr. During this time the heavier suspended particles and the floc settle out. This is called the sedimentation phase.

Following sedimentation the water flows through filters. The purpose of filtration is to remove remaining suspended particles. Normally, rapid sand filters are used. These filters pass about 3 gal/min/sq ft of filter area. Sometimes prepared anthracite coal is used as a filter medium, either by itself or added on a sand bed to increase the efficiency and maximum filter flow rate. The filter is cleaned periodically be reversing the flow through it to flush out accumulated matter. This process is called backwashing.

Throughout the treatment process additional chemicals are added. In addition to chemicals for disinfection, chlorine, potassium permanganate, or activated carbon may be added for taste-and-odor control. Lime, soda ash, caustic soda, metaphosphate or silicate may be added to control the corrosion characteristics of the water.

A treatment plant might also soften water. This is accomplished by one of two methods. All of the hardness can be removed by the ion exchange method. In this method hardness-causing calcium and magnesium salts are removed from the water by means of zeolite, which substitutes sodium ions. The other method is chemical softening, in which lime and soda ash are used to remove part of the hardness.

Water Quality Safeguards

The critical safeguards for a water system are continuous positive pressure in the mains, chlorine residuals, cross connection control, and frequent testing. Continuous positive pressure is necessary to prevent backsiphonage and the entry of contaminants. Continuous positive pressure is mostly the result of adequate supply and storage, proper design and correct operations to meet peak demands. However, loss of pressure may still occur following major water main breaks or with a large water demand during fires.

Chlorine residuals should exist throughout the system to prevent any bacterial contamination of the water. Where problem areas exist in the distribution system, frequent checking to ensure the presence of chlorine residual will help to safeguard the water.

Sampling of water for bacteriological analysis is required by the EPA. The number of samples and the frequency of the sampling is dependent upon the size of the population served, the initial water quality, and the bacteriological history of the system. When taking samples the operator should follow the instructions set forth by regulatory agencies to ensure that the sample taken is representative of the actual water in the system. The operator must ensure that the sampled water is not contaminated by the methods used in taking the sample.

Disinfection of Water

Water is disinfected in order to destroy, prevent, or minimize the growth of bacteria. Water may be pasteurized (by heat), subjected to ultraviolet radiation, or chemically treated to destroy or reduce the growth of bacteria. It may be chemically treated using, among other agents, chlorine, chlorine and ammonia, ozone, bromine, or iodine.

Chlorine (Cl) is the most commonly used disinfectant in the US. It can be used in solid, liquid, or gaseous form. At room temperature chlorine is a greenish yellow gas. The odor of chlorine, either as a gas or a liquid, is pungent and biting. Under pressure chlorine can be liquified at room temperature. This form is most used for shipping and storing. Liquid chlorine is amber in color.

Chlorine is supplied in different sizes of steel pressure containers, railway tank cars, or truck tanks. All steel tanks have fuse plugs that rupture when excessive pressure is put into the tank. At 160F (71C) the fuse plug will melt because the pressure in the cylinder increases with temperature. Chlorine may be drawn from a tank as either a liquid or gas. Because the liquid chlorine is concentrated and heavier, chlorine from the top of the tank would be gas and chlorine from the bottom of the tank would be in liquid form if under sufficient pressure.

Chlorine is regulated and injected as a gas into a stream of water, which in turn is fed as a solution into the water supply. (See Fig. III-2.)

Chlorine, in a steel cylinder, is under high pressure and is in liquid form. When the valve is open, the pressure on the released chlorine decreases and it becomes a gas. The release of this pressure lowers the temperature at the nozzle such that if the gas is released at too fast a rate, the nozzle will freeze shut.

Chlorine, as a liquid, can be withdrawn from the cylinder at a much faster rate, provided the pressure in the discharge line is approximately equal to that inside the cylinder. The liquid is piped to an evaporator where it is converted into a gas for injection. The evaporator is essentially a hot water bath container that provides sufficient heat to prevent the outlet nozzle from freezing when the liquid chlorine is converted to a gas at high rate.

Chlorine gas can be drawn at a rate of 40 lb/day from a 150-lb cylinder and about 400 lb/day from a 1-ton cylinder when room temperature is about 70F (21C). By contrast, liquid chlorine can be drawn from a 1-ton cylinder through an evaporator at a rate in excess of 8000 lb/day.

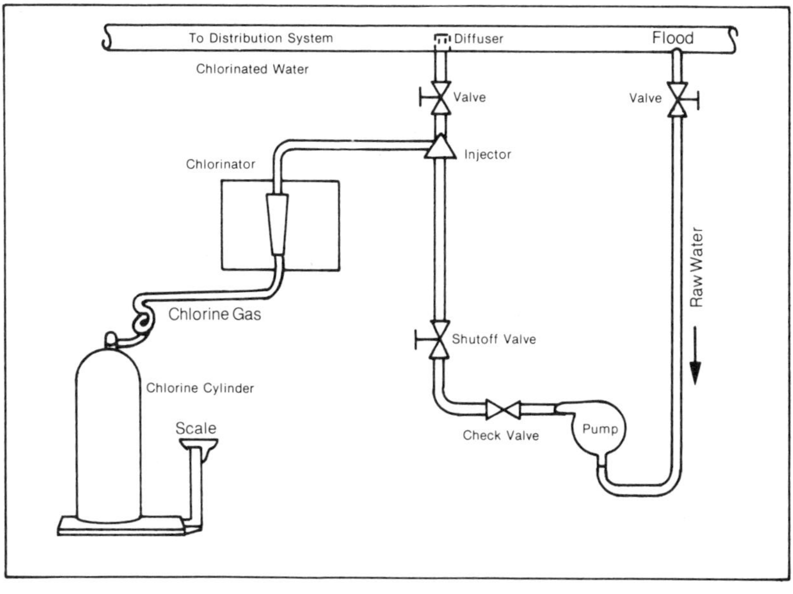

Fig. III-2. Typical Chlorination Installation

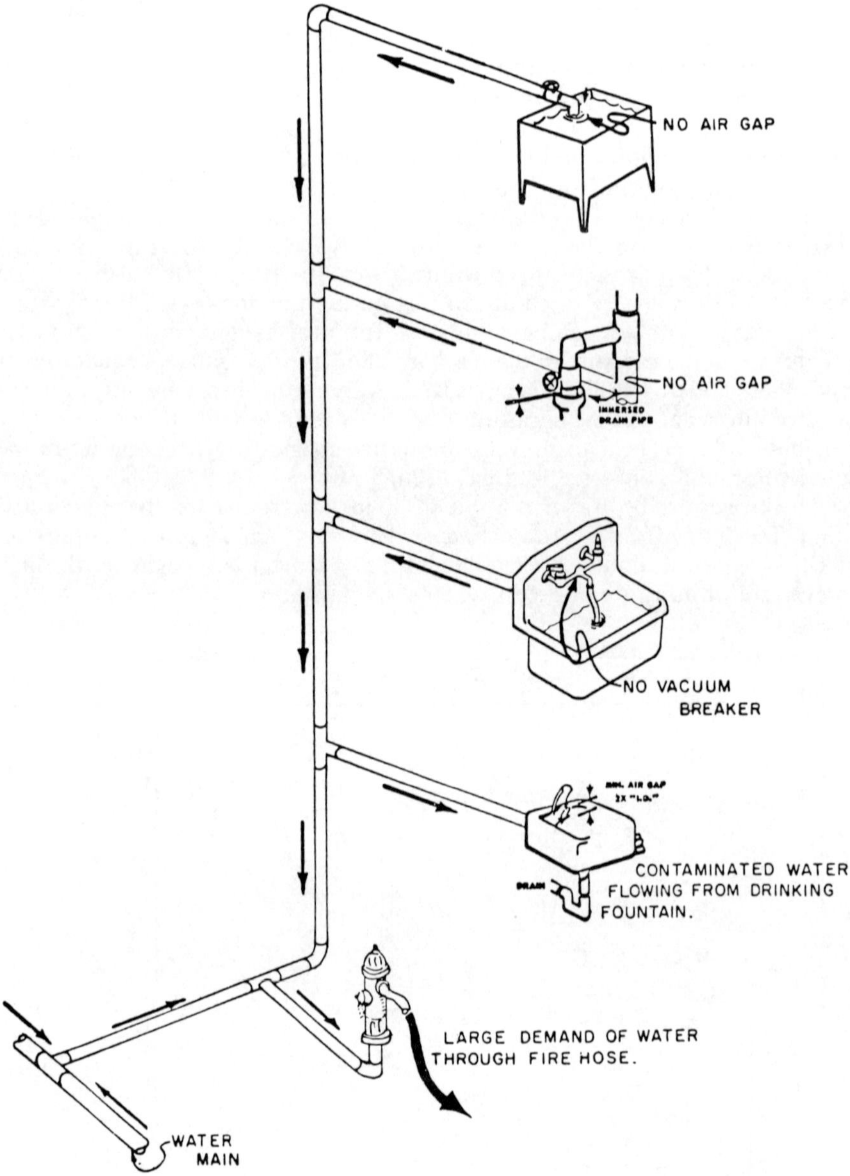

Fig. III-3. Example of Backflow Hazard

Time is required for the chlorine gas to spread throughout the water, and for disinfection to occur. Known as the contact time, it is a very significant variable in chlorination.

Chlorine is also available in a dry form as calcium hypochlorite or in solution form as sodium hypochlorite such as a household bleach. Hypochlorites are applied by mechanical feeders or in solution by hypochlorinators. Normally, hypochlorites are expressed as containing a certain percentage of available chlorine. This is very important in determining the application rate because if a 70 per cent hypochlorite is used, then all measurements for chlorine as a part per million of water must be adjusted for the 70 per cent factor, by dividing by 0.70. This will give the amount of hypochlorite to be applied for the desired chlorine.

Chlorine, when added to water, reacts not only with bacteria but also with other substances and organic matter. Because of this, sufficient chlorine must be added to leave an excess of chlorine in the water. This excess is called the chlorine residual and is expressed as ppm or mg/l. The total amount of chlorine added minus the chlorine residual gives the chlorine demand of the water. (See appendix.)

A chlorine residual is required throughout the distribution system and serves to safeguard the water quality. Chlorine residual may be of two different states—the free available state and the combined state. Residual chlorine in the free available state is an effective and rapid disinfectant. The most common combined state is chloramine, which is a combination of chlorine and ammonia. The combined state is slower acting as a disinfectant and for this reason residuals should be of the free available state. The free chlorine residual and the combined chlorine equal the total chlorine.

The orthotolidine (OT) and N, N-diethyl-p-phenylenediamine (DPD) tests are used to measure the chlorine residual. The procedures for these tests are given in the Appendix of this chapter.

Chlorine can be a very hazardous material. Safety aspects of chlorine handling and use will be covered in Chapters XIII and XIV. See Fig. III-2 for a typical chlorination installation.

Cross Connections

Cross connections are here defined as any connection between a system of potable water and another system of questionable or known poor quality. Examples of cross connections are those where the water supply has a connection to a sewer system, to sewer flushing chambers, hydrant drain lines connected to sewers, or direct connection of potable water piping to chemical pipes. Cross connection yielding back flow (Fig. III-3), or backsiphonage, represents the most critical factor in protecting any distribution system from contamination. Several devices are available and approved to prevent backflow and these will be discussed later in Chapters VII and IX. For the prevention of cross connection contamination every water utility should have an active cross connection prevention program that ensures that cross connections do not occur in its system.

Appendix

Residual Chlorine Tests

This test is used more than any other in water treatment. The word "residual" means remainder or that which is left. Thus, this test measures the amount of chlorine remaining in the sample at the time the test is made. For example, 2 ppm of chlorine may be added to a particular water. After a specific contact time, the test is made and if it indicates that 0.3 ppm chlorine is present, then 0.3 ppm is the chlorine residual. The difference between 2.0 pp and 0.3 ppm (1.7 ppm chlorine) was spent during the contact time in oxidizing the organic matter (bacteria, vegetable matter, sewage) in the water. This is known as the "chlorine demand" for the period of time. If 2 ppm of chlorine were added to distilled water, the chlorine residual test should show 2 ppm as there should be no demand because there is nothing in distilled water to be oxidized. It is important to note that, because of the chlorine demand, a chlorine residual test indicating sufficient chlorine at the plant does not necessarily mean that a test at the end of the main will indicate sufficient chlorine. (For example, iron bacteria or foreign matter introduced by backsiphonage could be in the lines exhausting the chlorine.)

Generally speaking, chlorine residuals are called "free chlorine residual" and "combined chlorine." The term "free chlorine residual" indicates that enough chlorine has been added and that some has remained. It could be said that some chlorine is left in the water and is free to do any further work that could be necessary. "Combined chlorine" indicates that there is enough chlorine present to be measured, but that it is combined with some other material. Combined chlorine is usually responsible for chlorinous tastes in water. Normally, additional chlorine will reduce complaints of tastes associated with chlorine.

Several chemicals can be used to test for the presence of chlorine in water. The most commonly used in field test kits is orthotolidine (OT). Oxidized manganese and iron, as well as chloramine, can react with orthotolidine to yield false free chlorine readings when the determination is performed carelessly. The orthotolidine arsenite method reduces these errors.

The DPD (N,N-diethyl-p-phenylenediamine) method is more accurate over wider temperature and pH ranges without the errors caused by interference. For these reasons, although the orthotolidine methods are still used, the DPD method is recommended and should replace completely the use of orthotolidine.

1. *Colorimetric testing using DPD method.*

   (a) Put free available chlorine reagent in sample bottle or tube. If tablet is used, it must be crushed with a glass tube.
   (b) Fill sample tube.
   (c) Cap tube and shake.
   (d) The color of violet indicates the amount of free available chlorine present. Compare with standard colors to determine amount of chlorine residual.

(e) Put total chlorine reagent in sample tube and shake.

(f) Wait 3 min then once again compare with standard color to determine amount of total chlorine.

(g) Subtract free available chlorine (chlorine residual) from total chlorine to determine combined chlorine.

2. *Colorimetric method using orthotolidine only.*

The following apparatus and reagents are needed:
A comparator with chlorine comparator color disc or a set of standard chlorine residual color bottles; an empty bottle that is the same size and diameter as the standard chlorine residual color bottles; and standard orthotolidine solution.

Procedure:

(a) Fill the sample bottle or tube with the water to be tested.

(b) Add 10 - 12 drops of orthotolidine indicator solution and mix. (droppers, which are marked, may be obtained so that the correct amount is indicated.)

(c) A yellowish color indicates the presence of residual chlorine. The intensity of color will indicate the amount of chlorine present.

(d) Rotate the comparator disc so that the color of the sample with orthotolidine added and a color on the disc match. The amount of chlorine present is read directly from the comparator. If color standards in bottles are used, compare the color of the sample with the color standards using a white background to secure best results. The amount of residual chlorine in the water is equal to the color standard that matches the color of the sample.

When free available chlorine, which is desirable, is present, the orthotolidine reacts with the clorine immediately and most of the color is produced within 5s. This immediate action is known as the flash test and the color produced within 5 s should be at least 80 per cent of the total color developed. When the color is produced slowly, requiring 3 - 5 min for the full intensity to develop, the indications are that the chlorine is combined with some other element. This combined chlorine may cause a chlorinous taste and may be eliminated by the addition of enough chlorine to obtain the flash color.

3. *Orthotolidine - arsenite (OTA) method.*

Nitrites, ferric iron compounds, manganese, lignocellulose, and algae are compounds or substances that could interfere with the chlorine residual test, causing false readings. Usually the effect of these compounds is to increase the apparent residual chlorine content of the sample. The orthotolidine-arsenite method was developed in order to decrease the effect of the interfering substances.

If interferences are present, additional tests are indicated to correct the values of total and free chlorine. The addition of sodium arsenite before the orthotolidine neutralizes all chlorine compounds in the water, and any color formed is caused by the reaction of interfering substances with the orthotolidine. The reading from the OTA test must then be subtracted from the original OT test results. The result of subtracting the result of the OTA test (interference) from the OT (total chlorine) is a true measure of the free chlorine residual present in the water.

Color variations of the DPD, OT, and OTA tests are often indicated by shining a light through the sample and reading the scale on a spectrophotometer.

Other methods for measuring chlorine in water include syringaldazine and leuco crystal violet test kits. Amperometric titrations and reverse titrations are generally used in laboratories.

Chapter IV
Operational Mathematics

Introduction

In the water utility industry, phrases such as gallons per day, cubic feet per second, lineal feet of trench, per cent grade, pounds of chlorine, and gallons per minute are part of the daily conversation. Much of the work requires arithmetic and some algebraic figuring.

Arithmetic is the science of positive real numbers. It deals with the number system and how it is used in the affairs of daily life. The concept of zero is thought by many to be one of the greatest inventions of modern civilization. Man enters this world with nothing and leaves it with nothing—zero. Hence, there must be a beginning and a true understanding of zero.

Most people are taught how to compute by the methods developed with the Hindu-Arabic system. To use these methods one must first memorize the basic combinations in each of four operations—addition, subtraction, multiplication, and division.

Mathematics is a subject in which one must apply himself. It is the one essential tool one must have; for without a basic knowledge of mathematics, how is an operator to understand the operation of his system? With a solid foundation in mathematics one could also get a better rating and progress in his job.

Basically, arithmetic involves only four categories.

1.	Addition	+ plus
2.	Subtraction	- minus
3.	Multiplication	x times
4.	Division	÷ into

Usually, the surest and safest method for solving arithmetic problems confronting water operators is the application of simple basic principles. Some computations are simplified by the use of arithmetical formulas, but caution should be exercised not to misuse any formula.

This chapter is designed for self teaching and as a refresher course. It covers, though briefly, addition, subtraction, multiplication, fractions, decimals, percentages, proportion, significant figures, algebra, geometry, units, chemical calculations, and hydraulics.

The plus sign (+). The plus sign is part of most numbers, but the sign itself is seldom used. The sign is taken for granted. When a number is written, it is a positive (+) number unless indicated otherwise. The plus sign is used to add one number to another, but one seldom indicates the + sign. It is usually assumed.

The minus sign (-). If one number is to be reduced by a certain amount, the result would be the first number minus the second number. This operation would be indicated by using the minus or negative (-) sign.

Negative numbers have equal "weight" but opposite direction to positive numbers. The minus sign is used in this case to give that number its identity, so the minus sign must be retained. For example, a temperature change from -20 to +20 would be a total change of +40.

The times sign (x). The times sign means to multiply and would be used to find out what the total number would be if many identical numbers were added together. One would not add 2 plus 2 plus 2; he would multiply 2 times 3.

The times sign is not used normally in the engineering field. Numbers to be multiplied are enclosed in parentheses and placed next to one another; example, (3) (2); or a dot can be used in place of the times sign; example, 3·2. Thus, there are in essence three times signs: 3 x 2, (3) (2), and 3·2.

The division sign (÷). Normally, in elementary use, the problem is taken from the books shown thus: 6÷2=. This would then be worked by short division, $2\underline{|6}$, or long division, $2\overline{)6}$. A method more commonly used would be with the problem written as a fraction 6/2. Thus, the division signs are 6÷2, $2\underline{|6}$, $2\overline{)6}$, and 6/2.

Other mathematical signs. There are also some basic symbols that do not instruct one to do an operation but are shorthand for written statements. Those most commonly used are

The equals sign	(=)	
The therefore symbol	(∴)	
The greater than sign	(>)	The point always faces
The less than sign	(<)	the smallest number.

Example: 12>4 means 12 is greater than 4.

Although the decimal system is a simple and xcellent system, one needs to pay special attention to the position of the decimal point. When working out answers to daily questions, it is essential to perform neat orderly work, preferably using square lined graph paper, which is most helpful in developing orderly methods and sketching.

Accuracy

Accuracy is a relative term. Many people list accuracies beyond those of the measuring devices. Stating on a sign that the population of a city is to the nearest one is absurd because the population changes by the hour and, in cases of large cities, by the minute. It is essential to be accurate to the nearest cent in money matters, and it is essential to be accurate to the nearest 1/10 in. when measuring fittings. But time and effort are wasted in unwarranted accuracy. For most calculations one can round off to the first three significant numbers: 72834.5 becomes 72800 and 0.0027346 becomes 0.00273.

It is of no advantage to be more accurate than the measurement warrants. Because most daily computations are connected with materials, time, and money, it is essential that one does not make computations that cannot be reasonably measured.

Mental Arithmetic

There are many techniques for mental arithmetic and these can be very valuable in the field or for checking. Most of them are based on a breakdown technique such as that illustrated in Table IV-1.

TABLE IV-1

Mental Arithmetic

Number	Using as a Multiplier	Using as a Divider
$5 = \frac{10}{2}$	$5 \times 362 = \frac{10}{2} \times 362 = \frac{3620}{2}$	$\frac{362}{5} = \frac{724}{10}$
	$= 1\,810$	$= 72.4$
$8 = 2 \times 2 \times 2$	$8 \times 92 = 2 \times 2 \times 2 \times 92$	$92 = \frac{92}{2 \times 2 \times 2} = \frac{46}{2 \times 2} = \frac{23}{2}$
	$= 2 \times 2 \times 184$ $= 2 \times 368$ $= 736$	$= 11½$
$9 = 10 - 1$	$9 \times 126 = (10 - 1)\,126$	$\frac{126}{9} = \frac{126}{3 \times 3}$
$= 3 \times 3$	$= 1\,260 - 126$	$= \frac{42}{3}$
	$= 1\,134$	$= 14$

Note that arithmetic accuracy is primarily in the decimal point. As a mental check, 5 x 362 from Table 1 is approximately 5 x 400 or 2000. Therefore, (∴), the answer is less than (<) 2000 or, 1810—not 18 000 and not 181. The approximate answer of 2000 is far more accurate than 18 100 or 181. Using approximations is always a good practice for checking the accuracy of one's answers, especially in relation to the placing of the decimal point.

In addition to the need for the use of arithmetic, it is also essential that one become familiar with some of the elements of other forms of mathematics.

Fractions

A fraction is an expression of unexecuted division. In itself, it is one or more of the equal parts of anything. The terms are the numerator and denominator. Similar fractions have the same denominator. A fraction whose numerator is smaller than its denominator is called a proper fraction, such as ½. If the numerator is larger than the denominator, it is called an improper fraction, such as $\frac{5}{2}$. A number such as $5\frac{3}{4}$ is called a mixed fraction. Remember: $\frac{\text{numerator}}{\text{denominator}}$.

Addition and Subtraction

When adding or subtracting fractions a common denominator must be used. The least common denominator is the smallest number into which each denominator can be divided one or more times with nothing left over.

Examples:

$$\frac{3}{4}+\frac{3}{8}=\frac{6}{8}+\frac{3}{8}=\frac{6+3}{8}=\frac{9}{8}=1\frac{1}{8}$$

$$\frac{2}{3}+\frac{3}{4}+\frac{7}{9}=\frac{24}{36}+\frac{27}{36}+\frac{28}{36}=\frac{24+27+28}{36}=\frac{79}{36}=2\frac{7}{36}$$

Multiplication

To multiply a fraction by a fraction, multiply the numerators for the new numerator and multiply the denominators for the new denominator.

$$\frac{2}{3}\times\frac{5}{7}=\frac{10}{21}$$

To multiply a fraction by an integer, multiply the numerator of the fraction by the integer and place over the same denominator. Or to consider it another way, an integer is a fraction whose denominator is one.

$$5\times\frac{3}{7}=\frac{15}{7}\quad\text{or}\quad\frac{5}{1}\times\frac{3}{7}=\frac{15}{7}$$

Division

To divide fractions, invert divisor then multiply.

$$\frac{2}{3}\div\frac{5}{7}=\frac{2}{3}\times\frac{7}{5}=\frac{14}{15}$$

$$5\div\frac{3}{7}=5\times\frac{7}{3}=\frac{35}{3}=11\frac{2}{3}$$

Cancellation

If the numerator and denominator of a fraction are multiplied or divided by the same number, the value of the fraction is not changed. This is the principle of cancellation.

If, however, the same number is added to or subtracted from both the numerator and denominator of a fraction, the value of the fraction is changed. For example,

$$\frac{3}{4}=\frac{3\times2}{4\times2}=\frac{6}{8}$$

but,

$$\frac{3}{4}\text{ is not the same as }\frac{3+2}{4+2}\text{, which is }\frac{5}{6}\text{ and not }\frac{6}{8}\text{ or }\frac{3}{4}.$$

Decimals

A decimal is a fractional part of a number expressed by using tenths (0.1), hundredths (0.01), or thousandths (0.001), etc.

Examples:

2/10 is expressed as 0.2
2/100 is expressed as 0.02
2/1000 is expressed as 0.002

When anything is divided into tenths or hundredths, the parts are called decimal divisions. One or more of the decimal divisions of a unit is called a decimal fraction.

$$0.1 = \frac{1}{10} \qquad 0.543 = \frac{543}{1\ 000} \qquad 0.00062 = \frac{62}{100\ 000}$$

Locating the decimal point

Annexing zeros to a decimal point does not alter its value for calculation. In 0.00624 the two zeros between the point and the six serve to locate the decimal point and have no other meaning.

The term "decimal places" means figures to the right of the decimal point in a number.

Additions and subtractions of decimal fractions

When placing numbers to be added that contain decimal fractions, place the numbers in such positions that all decimal points will be in the same vertical column. Should any of the numbers have no visible point, it should be understood that a decimal point may be placed just to the right of the last figure in that number. For example,

```
365.06
  3.10
 63.
 19.656
  0.002
-------
450.818
```

Any amount of zeros, if needed, may be placed to the right of a decimal point in a number without changing the value of that number. For example

```
365.060        19.000
  3.100       -16.167
 63.000       -------
 19.656         2.833
  0.002
-------
450.818
```

Multiplication of decimal fractions

When multiplying two numbers, either or both containing decimal fractions, multiply them as if no decimal fractions were present. Then count the decimal places in both the numbers, and that total will be the number of decimal places in the answer.

Example:

One gal of water weighs 8.34 lb
How much will 6.5 gal weigh?
8.34 lb x 6.5 = 54.210 lb

Division of decimal fractions

A method of dividing when the divisor contains a decimal is

$$\frac{8.35}{0.5} \text{ or } 0.5\overline{)8.35}$$

Move both decimals to the right until the divisor is a whole number. For example,

$$\frac{8.3_{\wedge}5}{5_{\wedge}} \text{ or } 0.5_{\wedge}\overline{)8.3_{\wedge}5}$$ ($\wedge$ indicates the new location of the decimal.)

This is equivalent to multiplying both numbers by a constant 10. The decimal in the answer will be directly above the new decimal in the dividend:

$$5\overline{)83.5}\quad 16.7$$

Common fractions into decimals

A common fraction is reduced to a decimal fraction by dividing the numerator by the denominator:

$$\frac{2}{5} = 0.4 \quad \frac{9}{16} = 0.5625 \quad \frac{2}{7} = 0.2857$$

Converting decimals to fractions

To change a decimal fraction to the nearest number of eighths, sixteenths, or thirty-seconds, divide by the decimal equivalent of 1/8, 1/16, or 1/32. So,

$$0.34375 \div 0.03125 \text{ (or } 1/32) = 11$$

Exponential Numbers

With the advent of atomic energy and the space age, the use of numbers with exponents is becoming more generally understood. It is also easier to multiply and divide large numbers or infinitely small numbers using exponents than zeros before or after the decimal point. For example, to multiply 10 million by 9 billion using the older notation:

$$10\,000\,000 \times 9\,000\,000\,000 = 90\,000\,000\,000\,000\,000$$

It is very easy to gain or lose a zero. Using the modern method or scientific notation,

$$10\,000\,000 = 10 \times 10^6$$
$$9\,000\,000\,000 = 9 \times 10^9$$
$$\text{and } 10 \times 10^6 \times 9 \times 10^9 = 90 \times 10^{15}$$

One simply multiplies the integers and adds the exponents.

To divide 10×10^6 by 9×10^9, divide the integers and subtract the exponents.

$$\frac{10 \times 10^6}{9 \times 10^9} = 1.11 \times 10^{-3}$$

Scientific notation

$$10^0 = 1$$
$$10^1 = 10$$
$$10^2 = 10 \times 10 = 100$$
$$10^6 = 10 \times 10 \times 10 \times 10 \times 10 \times 10 = 1\,000\,000$$
$$0.1 = 10^{-1}$$
$$0.01 = 10^{-2}$$
$$0.001 = 10^{-3}$$
$$0.0001 = 10^{-4}$$

Any number may be expressed as a number between 1 and 10 multiplied by some power of 10.

$$92\,000\,000 = 9.2 \times 10^7$$
$$349.2 = 3.492 \times 10^2$$
$$0.00498 = 4.98 \times 10^{-3}$$
$$0.00473 = 473 \times 10^{-5} = 4.73 \times 10^{-3}$$

The power of 10 used is equal to the number of places that the decimal point has been shifted; positive if moved to the left and negative if moved to the right.

Numbers in the scientific notation may be added or subtracted only when the powers of 10 are equal for both.

$$3.72 \times 10^3 + 8.56 \times 10^3 = (3.72 + 8.56) \times 10^3$$
$$= 12.28 \times 10^3$$
$$\text{or } 1.228 \times 10^4$$

$$4.53 \times 10^3 + 1.271 \times 10^4 =$$
$$4.53 \times 10^3 + 12.71 \times 10^3 =$$
$$17.24 \times 10^3 = 1.724 \times 10^4$$

Numbers in the scientific notation can be divided or multiplied.

$$\frac{10^m}{10^n} = 10^{m-n}$$

$$\frac{6 \times 10^4}{2 \times 10^2} = 3 \times 10^{4-2} = 3 \times 10^2$$

$$10^m \times 10^n = 10^{m+n}$$

$$(2 \times 10^2) \times (3 \times 10^3) = 2 \times 3 \times 10^2 \times 10^3$$
$$= 2 \times 3 \times 10^{(2+3)}$$
$$= 6 \times 10^5$$

In division, the main numbers (integers) are divided, and the power factors (exponents) are subtracted. In multiplication, the integers are multiplied and the exponents are added.

Percentage

Per cent literally means "by the hundred," or parts per 100 parts. Per cent is the numerator of a fraction whose denominator is always 100. "Per" means "divided by."

$$\frac{50}{100} = 50\% \qquad \frac{2.5}{100} = 2.5\%$$

What is 16 per cent of 80?

$$80 \times \frac{16}{100} = \frac{1\,280}{100} = 12.8 \text{ or approximately } 13.$$

Common fractions, decimal fractions, and percentages are related terms as shown in Table IV-2. When using per cent, it is important to recognize the base. If one has 8 g of a material that contains 3 g of sand, the ratio of the weight of the sand to the weight of the whole sample is 3 to 8. Here the base is the weight of the entire sample.

$$\frac{3}{8} = 0.375 = \frac{0.375}{1} = \frac{37.5}{100} = 37.5\%$$

Problems in Percentage

Problem: What weight of water is in a ton of sludge that is 90 per cent water?

TABLE IV-2
Relation Between Percentages and Fractional and Decimal Equivalents

Percentage	Common Fraction	Decimal Fraction	Scientific Notation
1.000	$\frac{1}{100}$	0.01	1×10^{-2}
14.000	$\frac{14}{100}$	0.14	1.4×10^{1}
360.000	$\frac{360}{100}$	3.60	3.6
0.002	$\frac{2}{100\ 000}$	0.00002	2×10^{-5}
0.0001	$\frac{1}{1\ 000\ 000}$	0.000001	1×10^{-6}

Ninety per cent means that 90/100 of any weight of sludge is water.

$$2\ 000 \text{ lb} \times \frac{90}{100} = 2\ 000 \text{ lb} \times 0.9 = 1\ 800 \text{ lb}$$

Problem: One lb of lime is stirred into 1 gal of water. What is the percent of lime in the slurry? (1 gal of water = 8.34 lb)

$$\frac{1}{1 + 8.34} \times 100 = 10.7\%$$

Problem: The chlorine demand of a liquid is 9 mg/l. After chlorination, the demand is 2 mg/l. What is the percentage of satisfaction of the demand?

$$9\text{-}2 = \text{part of demand satisfied}$$

$$\frac{9-2}{9} = \frac{7}{9} = 0.77$$

$$0.77 \times 100 = 77 \text{ per cent satisfaction}$$

Problem: What weight of water and alum are needed to make 5 000 lb of a 5 per cent solution?

$$5\ 000 \times \frac{5}{100} = 250 \text{ lb of alum}$$

$$5\ 000 - 250 = 4\ 750 \text{ lb of water}$$

$$\text{Total} = 5\ 000 \text{ lb (water + alum)}$$

Observe that a 5 per cent solution means 5 lb of alum in 100 lb of solution. Both the alum and the solution are expressed in the same units. This is, in general, the case when per cent is being considered.

To arrive at the gallons of water required to compose the 5 per cent solution, it is necessary to divide pounds of water by the factor 8.34 (the weight of 1 gal of water).

$$\frac{4\ 750 \text{ lb}}{8.34 \text{ lb/gal}} = 570 \text{ gal}$$

Problem: If 3.75 lb of HTH (high test hypochlorite) containing 70 per cent available chlorine is dissolved in 30 gal of water, what is the percentage of available chlorine in the solution?

$$3.75 \text{ lb} \times 0.70 = 2.63 \text{ lb available chlorine}$$

$$\frac{3.75 \times 0.70}{30 \times 8.34} = \frac{2.63}{250} = 0.0105 \text{ lb available chlorine in 1 lb of solution}$$

$$100 \times 0.0105 = 1.05\% \text{ available chlorine (or 10 500 ppm of chlorine)}$$

In dilute solutions, the weight of water may be taken as equal to the weight of the soultion.

Problem: Suspended solids in the influent are 60 mg/l and in the effluent 20 mg/l. What is the per cent efficiency of the settling basin?

$$60 \text{ mg/l} - 20 \text{ mg/l} = 40 \text{ mg/l removed}$$

$$\frac{40}{60} = 0.667$$

$$0.667 \times 100 = 66.7 \text{ efficiency}$$

Significant Figures

Two kinds of numbers are used in practical calculations. They are absolutes and measurements. An "absolute" expresses the value of something exactly. "Measurements" are the readings of scales, balances, meters, gages, and manometers. Every device for measuring has a threshold or sensitivity limit below which a change in the item measured will not be detectable. Every number representing a measurement is thus inexact or incomplete.

If a meter shows a reading as in Fig. 1, the reading one would report would be $7._2$ or 7 as the second figure (2) must be estimated by eye and the only figure one can be sure of is the 7. When a figure is estimated as 7.2, it is customary to write it as $7._2$.

When measurements are combined by arithmetical operations, care is needed to avoid results that are absurd, meaningless, or actually fictitious.

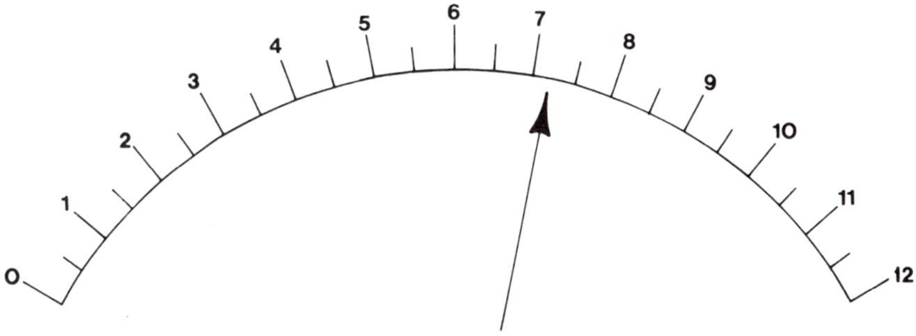

Fig. IV-1. Estimated Measurement

Determining significant figures

A significant figure in a number expresses the number of units in the place in which it is located. The decimal point has nothing to do with significant figures. Zero may or may not be significant. In 0.0006507, the zeros preceding the 6 are not significant, but the zero between the 5 and 7 is significant.

The number of significant figures in a measurement is determined by the instrument and the conditions under which it is used. If the instrument is properly constructed, in good adjustment, and used by a trained observer, its readings will be correct "as far as they go." What does "as far as they go" mean? If one has a balance that will not indicate differences in mass of less than 0.1 g, then any object weighed on that balance will be unknown beyond the first decimal place. If an object weighs somewhere between 12.8 and 12.9 and is recorded as 12.85, the weight of the object is really some absolute number between 12.8 and 12.9; that is, 12.85 ± 0.05.

Significant figures and rounding off

As numbers are usually written, zero may have two meanings. It is used as a significant figure and also to locate the decimal point. In 22 400 as a measurement, one cannot be sure which meaning is to be given to the zeros. It would be better to write it as 2.24×10^4. This clearly indicates only three significant figures. The answer should have no more significant figures than the least number of significant figures in its parts.

Problem: The following measurements are to be added: 22 400, 3 642.2, and 23.24. The zeros in 22 400 are not significant; therefore, only three significant figues are meaningful.

```
 22 400
  3 642.2
     23.24
 ---------
 26 065.44–should be rounded off to 26 100
```

Problem: Add 22.4, 3.675, and 0.8842.

```
 22.4
  3.675
  0.8842
 -------
 25.9592
```

or round off and rewrite

```
 22.4
  3.7
  0.9
 ----
```

Because the second decimal place is not significant, the answer is 27.0.

Proportion

The indicated division of two pure numbers is a ratio. The ratio of 3 to 4 is usually expressed as 3:4 or sometimes as 3/4. A proportion is an equality of two ratios and is written in the form $a{:}b = c{:}d$ or $a/b = c/d$ and is expressed as a is to b as c is to d. The terms a and d are the extremes and b and c are the means. The product of the extremes in a true proportion must always equal the product of the means (sometimes called cross multiplication).

Therefore, from the proportion

$$a/b = c/d$$

one gets

$$a \times d = c \times b$$

(extremes) (means)

To find any one of the values when the other three are known, proceed as follows:

$$a = \frac{c \times b}{d} \quad (1)$$

$$d = \frac{c \times b}{a} \quad (2)$$

$$\frac{a \times d}{c} = b \quad (3)$$

$$\frac{a \times d}{b} = c \quad (4)$$

The rule to remember is that when transferring the known value in the extremes or the means from one side of the equation to the other when solving for the unknown value, it is necessary to divide the product of the means or extremes (as the case may be) by the value transferred.

Another use of cross multiplication is to determine if two ratios are in porportion.

Examples:

1 —— 2
2 —— 4
(2)(2) = 4 = (4)(1) = 4

3 —— 42
7 —— 98
(7)(42) = 294 = (98)(3) = 294

Problem: 4.7 lb of a chlorine compound are dissolved in 45 gal of water. A second solution with 32 gal of water has 3.1 lb of the chlorine compound dissolved in it. Do the two solutions have a similar concentration of the chlorine compounds?

When one is considering a concentration, he is essentially concerned with a ratio. And, therefore, when comparing two ratios:

4.7 —— 3.1
45 —— 32
(45)(3.1) - 139.5 ≠ (32)(4.7) = 150.4

Because the cross multiplication does not yield similar products, one may assume that the ratios are different and therefore the concentrations are also different.

Direct and inverse proportion

Two varying quantities are directly proportional if the ratio of corresponding values is constant.

Problem: If 3.75 lb of HTH are added to 30 gal of water to make a solution of given strength, how many pounds will have to be added to 147 gal to make a solution of the same concentration?

$$3.75/30 = X/147$$

multiplying the extremes and the means yields

$$3.75 \times 147 = 30X$$

and solving for X, the unknown portion,

$$\frac{3.75 \times 147}{30} = X = 18.4 \text{ lb}$$

Two varying quantities are inversely proportional if their product is constant.

Units of Measurement

Before one can solve mathematical problems, it is necessary that he gain a clear understanding of the units of measurement with which he is working. Basically, there are two systems of measurement: (1) the English system, and (2) the metric system. Table IV-3 presents a list of conversion factors and illustrates the relationships between the two systems of measurement and abbreviations used in each.

TABLE IV-3

Selected Units of Measurement

English System		*Metric System*
	Linear	
1 inch (in.) (")		= 2.54 centimeters (cm)
1 foot (ft) (') = 12"		= 30.48 cm
1 yard (yd) = 36" = 3'		= 91.44 cm
39.37" = 3.28" = 1.094 yd		= 1 meter (m)
1 mile = 5280'		= 1.61 kilometers (km)
	Area	
1 square inch (sq in.) (in.^{2})		= 6.45 sq cm
1 square foot (sq ft) (ft^2)	= 144 sq in.	
1 square yard (sq yd) (yd^2)	= 9 sq ft	
1 acre (ac)	= 43 560 sq ft	
1 square mile	= 640 ac	= 2.59 sq kilometers
	Volume and Capacity	
1 cubic inch (cu in.) (in.^{3})		= 16.39 cubic centimeters (cc)
1 cubic foot (cu ft) (ft^3)	= 1 728 cu in.	
	= 7.48 gallons	
1 gallon (gal)	= 4 quarts (qt)	
	= 8 pints (pt)	
	= 128 fluid oz	= 3.785 liters (l)
		1 liter = 1 000 milliliters (ml)
		1 000 cubic centimeters (cc)
	Weight	
1 pound (lb) (#) = 16 ounces (oz)		= 453 grams (g)
		= 7 000 grains (gr)
2.205 pounds		= 1 kilogram (kg)
1 cu ft water	= 62.4 lb	1 liter water = 1 000 g = 1 kg
1 gallon water	= 8.33 lb	1 milliliter water = 1 g
	Rate of Flow	
	Metric system not generally used; the most common units are:	
Gallons per minute (gpm)		1 gpm = 60 gph
Gallons per hour (gph)		1 gpm = 1 440 gpd
Gallons per day (gpd)		1 mgd = 694 gpm = 41 440 gph = 1.547 cfs
Million gallons per day (mgd)		1 cfs = 60 cfm = 3 600 cfh
Cubic feet per second (cfs)		= 86 400 cfd
		= 646 300 gpd = 0.6463 mgd

Metric-to-metric conversions

Before the performing of any conversions, it is essential to be acquainted with the six common metric prefixes crucial to those conversions. These prefixes should be learned before conversions are attempted.

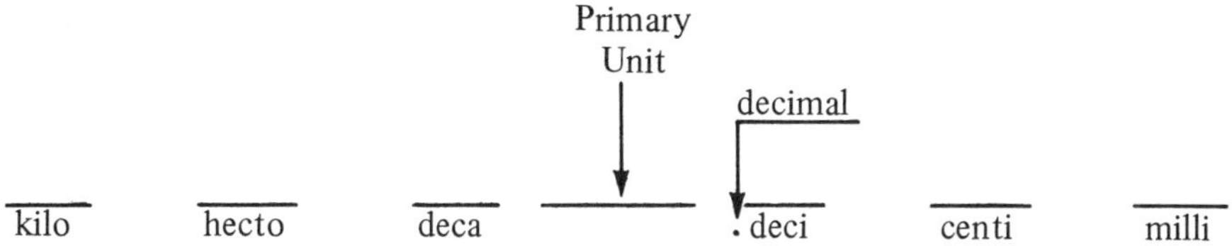

The metric system is based on units (powers) of ten just as is the decimal system. There are many methods of conversion, some of which are more difficult than others, but the student should find this method of conversion to be one of the easier methods.

In conversions of linear measurement, capacity measurement, and weight measurement, each place value will represent one decimal move. This system of conversion is best explained by example.

Example 1: [Metric to metric conversions]

Convert 1m to decimeters:

Converting from meters to decimeters, move one place to the right. Thus, one moves the decimal place from its present position one place to the right; 1.0 m = 10 dm.

Example 2: [Metric to metric conversions]

Convert 1g to (a) decigrams; (b) centigrams; and (c) milligrams.

deci centi milli

1 gram

(a) 1 place to the right:
1 g = 10 dg

(b) 2 places to the right:
1 g = 100 cg

(c) 3 places to the right:
1 g = 1 000 mg

This sytem of conversion will apply whether converting from the primary unit, as in examples 1 and 2 or whether converting from any other unit.

The Symbol π

The symbol π is the Greek letter pi, pronounced "pie." In mathematics, π is used as a constant in solving problems relating to a circle. π is the ratio between the circumference and the diameter of a circle and is equal to 3.1416 or about 22/7. For all practical purposes, one can use 3.14 as the value for π.

In some calculations it is convenient to use the value 0.785 which is $\pi/4$.

Abstract and Concrete Numbers

An abstract or pure number is used without reference to anything in particular. A concrete number applies to a particular item and is the product of an abstract number and a physical unit.

Examples of physical units are 1 ft; 1 acre; 1 gal.

Rule: Concrete numbers can be added or subtracted only when they possess the same physical unit.

Rule: Concrete numbers having the same or different physical units can be multiplied and, when this mutiplication takes place, the physical units must also be multiplied, thereby creating a new physical unit.

$$\begin{aligned} 6 \text{ ft} \times 7 \text{ ft} &= 6 \text{ (1 ft)} \times 7 \text{ (1 ft)} \\ &= 6 \times 7 \times \text{(1 ft)} \times \text{(1 ft)} \\ &= 42 \times \text{ft} \times \text{ft} \\ &= 42 \text{ ft}^2 \end{aligned}$$

1 ft^2 or 1 sq ft is a new physical unit. Similarly, 4 (1 ton) x 3 (1 mile) = 12 (1 ton) x (1 mile) = 12 ton-miles.

A ton-mile is a definite physical unit and represents something done, associated with carrying a ton through a distance of 1 mile.

5 (1 man) x 4 (1 hour) = 20 (man-hours)

1 000 (1 person) x 365 (1 day) = 365 000 person (capita) days

Rule: When concrete numbers are divided, the units are also divided. A unit in the numerator cancels a like unit in the denominator.

In general, when units are divided a new unit is created.

$$\frac{20 \text{ mi}}{2 \text{ hr}} = \frac{20}{2}\left[\frac{\text{(1 mi)}}{\text{(1 hr)}}\right] = 10\left[\frac{\text{(1 mi)}}{\text{(1 hr)}}\right]$$

or 10 mph.

Important Uses of Units

The result of every calculation with concrete numbers will generally be associated with a unit that gives meaning to the numerical answer. The numerical answer should always be accompanied by its unit. Columns of data must always have the units expressed at the top of the column. An important use for units is to determine how figures are to be combined to give the desired answer and to check the correctness of formulas for omitted quantities and incorrect units. Both sides of an equation must reduce to the same unit.

When discussing addition and subtraction, it was pointed out that entries must be in the same units. If given the following measurements to add, how would one do it? 2 feet + 10 miles + 60 meters + 100 yards = ? First change them all to the same units. For example, change them all to feet.

$$\begin{aligned} 2 \text{ ft} &= 2 \text{ ft} \times 1 &&= 2 \text{ ft} \\ 10 \text{ mi} &= 10 \text{ miles} \times (5\ 280 \text{ ft/mi}) &&= 52\ 800 \text{ ft} \\ 60 \text{ m} &= 60 \text{ m} \times (3.28 \text{ ft/m}) &&= 197 \text{ ft} \\ 100 \text{ yd} &= 100 \text{ yd} \times (3 \text{ ft/yd}) &&= \underline{300 \text{ ft}} \\ & && 53\ 299 \text{ ft} \end{aligned}$$

This total could be rounded off to 53 300 ft.

Calculating units

Sometimes a desired unit is not in the given list. If the desired unit is not present, it may usually be calculated with little difficulty.

The calculation involves two principles: (1) units may be cancelled and (2) any quantity divided by its equivalent equals unity or one.

For example, to obtain the number of centimeters in 1 ft, this calculation resolves itself simply into multiplying the number of centimeters in 1 in. by the number of inches in 1 ft.

$$12 \text{ in.} \times 2.54 \text{ cm/in.} = 30.48 \text{ cm}$$

Problem: What quantity of water weighs 1 lb? If proportion, explained earlier, is used to convert the value of 8.34 lb/gal to gal/lb then

$$\begin{aligned} 8.34 \text{ lb/1gal} &= 1 \text{ lb/X gal} \\ 8.34 \text{ X} &= (1)(1) \\ \text{X} &= 0.12 \text{ gal (not quite pint)} \end{aligned}$$

Problem: 1 oz/gal is equal (a) to how many lb/cu ft and (b) to how many g/l?

$$\text{(a) } 1\frac{\text{oz}}{\text{gal}} = 1\frac{\text{oz}}{\text{gal}} \times \frac{7.48 \text{ gal}}{1 \text{ cu ft}} \times \frac{1 \text{ lb}}{16 \text{ oz}} = \frac{7.48 \text{ lb}}{16 \text{ cu ft}} = 0.468\frac{\text{lb}}{\text{cu ft}}$$

$$\text{(b) } 1\frac{\text{oz}}{\text{gal}} = 1\frac{\text{oz}}{\text{gal}} \times \frac{1 \text{ gal}}{3.785 \text{ l}} \times \frac{28.3 \text{ g}}{1 \text{ oz}} = \frac{2.38 \text{ g}}{3.786 \text{ l}} = 7.48\frac{\text{g}}{\text{l}}$$

What Is Area?

Area is a measurement in two directions. When measuring length, use inches, feet, yards, or miles, but area is expressed in square inches, square feet, square yards, and square miles. The term "square" has a simple meaning in arithmetic–it means a number multiplied by itself. Knowing that the measurements of a desk are 5 ft by 3 ft, one can find the area by multiplying the length by the width. In other words, the area of the desk top is 5 ft x 3 ft = 15 ft^2 or 15 sq ft.

Figure IV-2 shows typical plane forms and the formula for calculating areas.

One thing to remember is that when working problems of measurement: not only multiply or divide the numbers, but also the terms or units such as feet, gallons, pounds, or whatever they may be. In the previous example the number 5 was multiplied by 3 and the product was 15. Also, feet were multiplied and the result was square feet ($feet^2$). Combining the numbers with the units produced an answer of 15 ft^2 or 15 sq ft. The answer when calculating area is always written in square inches, square feet, square yards, or square miles.

Problem: A road patch is 4 yd by 50 ft. What is the area?

$$4 \text{ yd} \times 3 \text{ ft/yd} \times 50 \text{ ft} = 600 \text{ sq ft}$$

Problem: A meter shop was measured to be 6 ft 9 in wide by 22 ft 4 in long. What is the area of the shop?

$$6\ 3/4 \text{ ft} \times 22\ 1/3 \text{ ft} = 6.75 \times 22.33 \text{ sq ft} = 150.7275 \text{ sq ft or } 151 \text{ sq ft}$$

One is also able to measure many various shapes. The fact that the object is measured by square inches (sq in.), square feet (sq ft), or square yards (sq yd) does

Rectangle:

The area, A, of a rectangle is equal to the product of its length and its width.

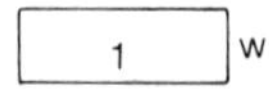

$A = lw$

Square:

The area, A, of a square is equal to the square of one of its sides.

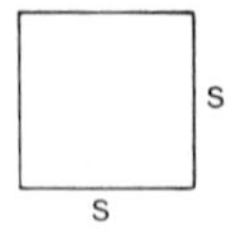

$A = s^2$

Perimeter: The perimeter, P, of a square may be found by multiplying the length of one side by four.

$P = 4s$

Parallelogram:

The area, A, of a parallelogram is equal to the product of its base and altitude.

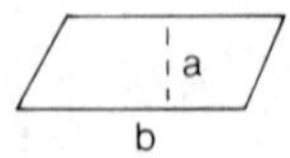

$A = ba$

Trapezoid:

The area, A, of a trapezoid is equal to one-half the sum of its base and top multiplied by its height.

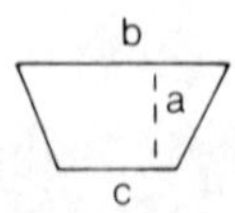

$A = \frac{1}{2}(b + c)a$

Lines:

Any two lines cut by three or more parallel lines are divided into proportional segments.

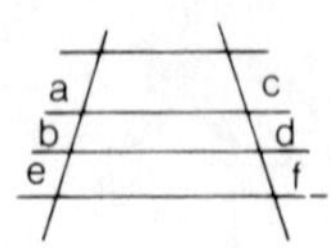

$a : b = c : d$

$b : e = d : f$

Triangle:

The area, A, of a triangle is equal to one-half the product of its base and its altitude.

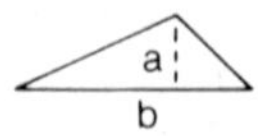

$A = \frac{1}{2}ba$

The square of the hypotenuse is equal to the sum of the squares of the other two sides for any right triangle.

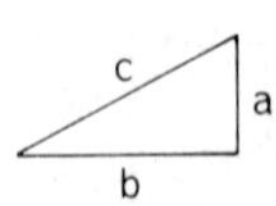

$a^2 + b^2 = c^2$

A line parallel to the base of a triangle divides the other two sides proportionally.

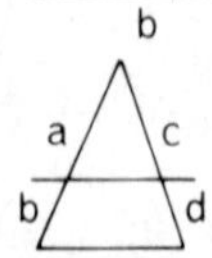

$a : b = c : d$

$(a + b) : b = (c + d) : d$

$(a + b) : a = (c + d) : c$

Fig. IV-2. Typical Shapes and Their Area Calculations

not mean that the object must be a square in order to measure it. Any shape can be measured. An acre is also a square measurement. It is equal to 43 560 sq ft.

Though the shapes differ in the following examples, the total surface area is the same. Most problems of surface area calculations involve one or more combinations of rectangles, triangles or circles (see Fig. IV-3A).

Example: (Area of combined shapes)

A channel has dimensions as shown in Fig. IV-3B. Calculate the total cross sec-

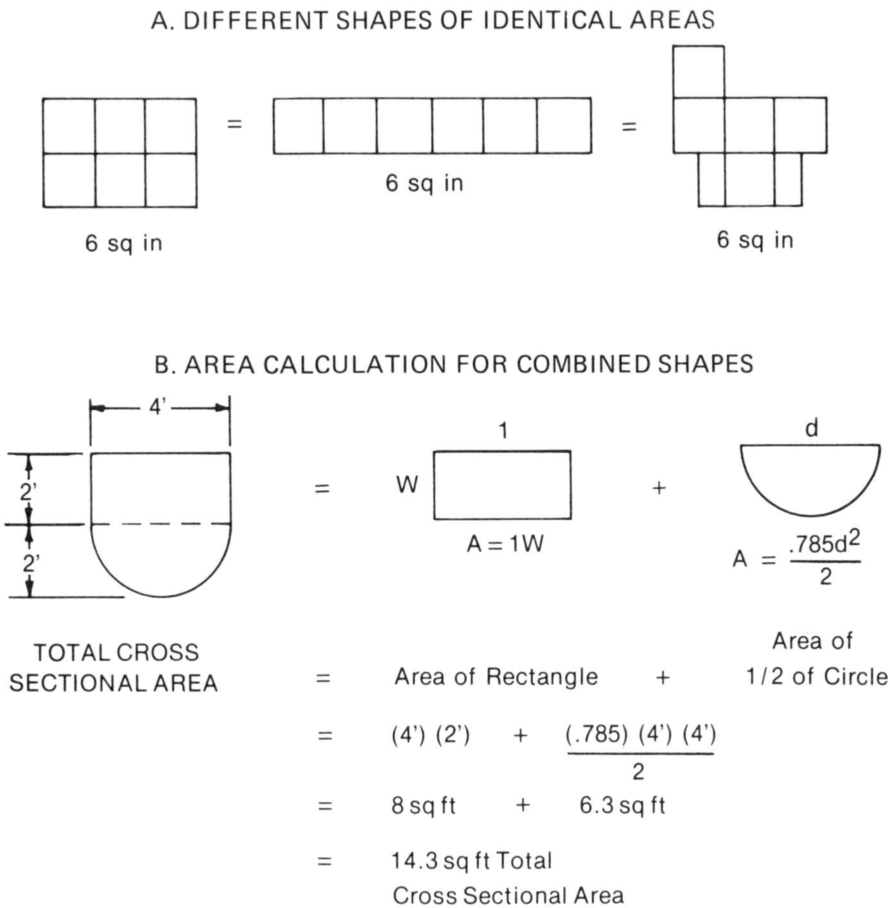

Fig. IV-3. Area of Combined Shapes

tional area of the channel.

What Is Volume?

Volume is measurement in three directions. When measuring length in the English system, use inches, feet, yards, or miles, and when measuring area, use square inches, square feet, square yards, or square miles. One measuring volume expresses it in cubic feet, cubic inches, or cubic yards. The term "cubic" in arithmetic is expressed as "cubic unit of length" or "volume." The cube of a number is that number multiplied by itself three times and is written with a 3 above the right side of the number that is cubed.

$$5 \times 5 \times 5 = 5^3 = 125$$

or, in the same number one can multiply the units of measurement, 5 ft by 5 ft by 5 ft = $(5\text{ ft})^3$ or 125 ft^3 or 125 cu ft.

It is important to remember that each dimension should be expressed as a single unit–i.e., feet, inches, meters, gallons–and not a mixed unit–i.e., feet and inches or gallons and quarts. Most often each of the units being multiplied will be identical so that the answer will be of a form $(\text{unit})^3$. Sometimes as "acre-ft," the volume is expressed in mixed units.

Example: What is the volume of a rectangular tank 45 ft long by 12 ft wide by 8 ft deep?

$$45\text{ ft} \times 12\text{ ft} \times 8\text{ ft} = 4\,320\text{ cu ft}$$

Circle:
The area, A, of a circle is equal to π multiplied by the radius squared, or 0.785 times the diameter squared.

$A = \pi r^2$ or $0.785 d^2$

Sphere:
The area, A, of a sphere equals the product of 4π and the radius squared.

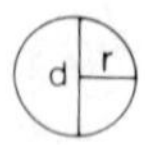

$A = 4\pi r^2$

The volume, V, of a sphere is equal to $\frac{\pi}{6}$ times the diameter cubed.

$V = \frac{\pi}{6} d^3$

Cylinder:
The lateral area, A, of a cylinder equals 2π times the product of its radius and height, L.

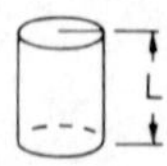

$A = 2\pi rL$

$V = \pi r^2 L$ or $0.785 d^2 L$

The volume, V, of a cylinder equals π times the product of the radius squared and the height.

Cone:
The volume, V, of a cone equals one-third π times the radius squared times the height.

$V = \frac{\pi}{3} r^2 a$ or $\frac{\pi}{12} d^2 a$

$V = \frac{1}{3}$ (area of base) a
(same as pyramid)

Pyramid:
The volume, V, of a pyramid equals 1/3 times the area of the base multiplied by the height, a.

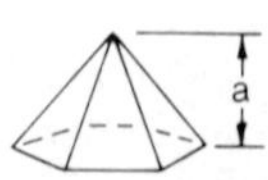

$V = \frac{1}{3}$ (area of base) a
(Same as cone)

Frustum of Cone or Pyramid:
The volume, V, of the frustum of a cone or pyramid equals 1/3 *height* a, times $A_1 + A_2 + \sqrt{A_1 A_2}$ where A_1 and A_2 are the area of the base and top.

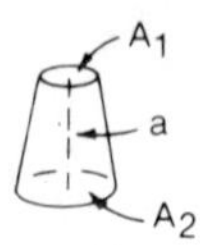

$V = \frac{a}{3} (A_1 + A_2 + \sqrt{A_1 A_2})$

Parallelepiped:
The volume, V, of a parallelepiped (rectangular solid) such as a cube or rectangular basin may be found by multiplying its length by its width by its height.

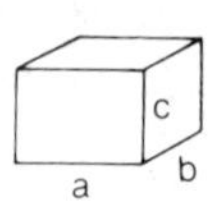

$V = abc$

Fig. IV-4. Typical Objects and the Formulas for Their Volumes

Example: What is the volume of the tank previously mentioned—20 ft 4 in. long by 10 ft 3 in. wide and 8 ft 6 in. deep?

20 ft 4 in. x 10 ft 3 in. x 8 ft 6 in. or
20 1/3 ft x 10 1/4 ft x 8 1/2 ft or
20.333 ft x 10.25 ft x 8.5 ft = 1 771.51 cu ft or 1 772 cu ft

Formulas for common shapes are given in Fig. IV-4.

For a shape of constant uniform cross section

Volume = Area of the Surface x Third Dimension

The "area of the surface" is the area of the representative surface, the side that gives the object its basic shape. In water utility calculations, these surface areas are most often rectangles, triangles, circles, or a combination of these as depicted in Fig. IV-5.

Once it is determined which surface of an object defines its basic shape, calculation of volume becomes a simple matter. The concept of the "third dimension" will be covered more fully later in this chapter.

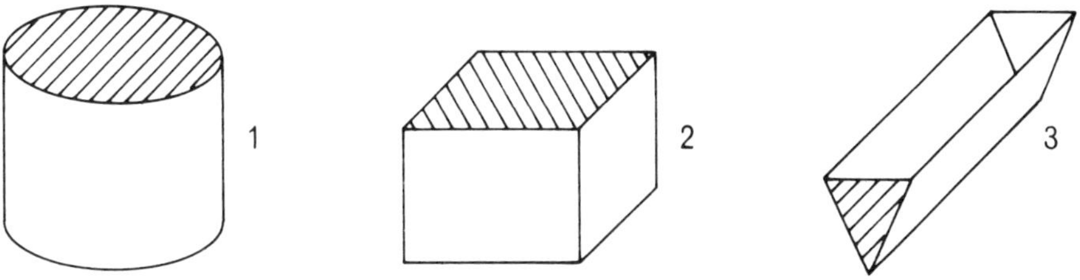

Fig. IV-5. "Area of Surface" for Cylindrical, Rectangular, Triangular

Rates

The concept of rate is very important in all practical affairs. A rate is a number multiplied by a unit divided by another different unit. Velocity, for example, measures rate of flow.

What is velocity?

Velocity is defined as the change of position per unit of time or as the rate of change of position. Velocity is speed. When a car travels 50 mph, the velocity of the car is 50 mph. Thus it travels 50 mi in 1 hr. Velocity of flow of water is no different except that it is expressed in different terms, usually feet per minute (fpm) or feet per second (fps). To find the velocity of flow, divide the distance traveled by the time spent traveling that distance. Thus, if an object in a flowing stream moves 30 ft in 1 min, the velocity is 30 fpm. To convert this velocity in fps

$$30 \text{ fmp} = \frac{30 \text{ ft}}{1 \text{ min}} = \frac{30 \text{ ft}}{60 \text{ s}} = \frac{0.5 \text{ ft}}{\text{s}} = 0.5 \text{ fps}$$

What is rate of flow?

It has been stated that velocity is the change of position per unit of time. Applied to a stream of water it is the speed of that stream of water. It was determined how fast a stream of water was flowing at a certain spot, or in other words, the change of position of an arbitrary "point" within the water. However, it was not determined how much water was flowing in the stream past that spot in a unit of time. This is the rate of flow, the volume of water flowing past a given spot in a unit of time. Rate of flow may be expressed as gallons per unit of time. Pump capacities are usually expressed in gallons per minute (gpm) and report sheets on plant operations ask for rate of flow in millions of gallons per day (mgd).

In the foregoing example, it was noted that the water moved 30 ft in 1 min, or that its velocity was 0.5 fps. Now suppose that the stream is 1 ft deep and 2 ft wide. In other words, the area of the cross section of the stream is 1 ft by 2 ft = 2 ft^2 or 2 sq ft. It was also determined that its velocity was 0.5 fps. To find the rate of flow of the stream, merely multiply the cross-sectional area of the stream by its velocity. The equation is $Q = A\,V$.

$$Q \text{ (rate of flow)} = A \text{ (area)} \times V \text{ (velocity)} = 2\ \text{ft}^2 \times 0.5\frac{\text{ft}}{\text{s}} = 1\frac{\text{ft}^3}{\text{s}} = 1\ \text{ft}^3/\text{s} =$$

$$1 \text{ cfs (cubic feet per second)}$$

This is how to find the rate of flow of a stream. Suppose one wanted to find the flow of a pump. He could do it in the same way if he knew the area of the cross section and the velocity of flow in the pipe at the pump. However, there is usually an easier way to find the capacity or rate of flow of a pump. Usually, one can measure the amount of water pumped in a certain period of time and from that determine the rate of flow of the pump. For example, pumping into an empty rectangular tank 30 ft long by 10 ft wide by 6 ft deep, one finds that it takes the pump 1 hr 15 min to fill the tank. The pumping rate or rate of flow of the pump is then calculated as follows:

To find the volume pumped—

$$30\ \text{ft} \times 10\ \text{ft} \times 60\ \text{ft} \times \frac{7.5}{\text{ft}^3}\ \text{gal}$$

first multiply the numbers

$$30 \times 10 \times 60 \times 7.5 = 13\,500$$

then multiply and divide the units

$$\text{ft} \times \text{ft} \times \text{ft} \times \frac{\text{gal}}{\text{ft}^3} = \text{ft}^3 \times \frac{\text{gal}}{\text{ft}^3} = \text{gal}$$

or do both operations in one step

$$13\,500\ \text{ft}^3 \times \frac{\text{gal}}{\text{ft}^3} = 13\,500\ \text{gal}$$

The volume pumped was 13 500 gal. This was pumped in 1 hr 15 min, or 75 min. The rate of flow or rate of pumping was 13 500 gal/75 min. It was stated that pump capacities were expressed in gallons per minute, to convert the rate of flow to gallons per minute

$$13\,500 \text{ gal per } 75 \text{ min} = \frac{13\,500 \text{ gal}}{75 \text{ min}} = 180 \text{ gpm}$$

This entire computation can be performed in one step:

$$30 \text{ ft} \times 10 \text{ ft} \times 60 \text{ ft} \times \frac{7.5 \text{ gal}}{\text{ft}^3 \times 75 \text{ min}} = 180 \text{ gal/min} = 180 \text{ gpm}$$

To convert this into million gallons per day, there are 60 min in an hour and 24 hr in a day, or 1 440 min in a day. If the pump will pump 180 gpm, in one day it will pump

$$\frac{180 \text{ gal}}{\text{min}} \times \frac{1\,440 \text{ min}}{\text{day}} = \frac{259\,200 \text{ gal}}{\text{day}} = 259\,200 \text{ gpd} = 0.2592 \text{ mgd}$$

Flow Conversions

The relationship among various water flow units is shown in IV-6.

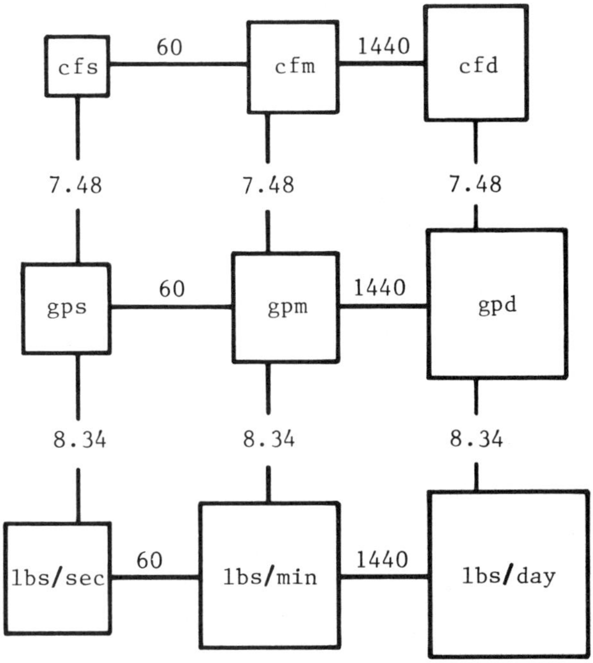

cfs = cubic feet per second
cfm = cubic feet per minute
cfd = cubic feet per day

gps = gallons per second
gpm = gallons per minute
gpd = gallons per day

Fig. IV-6. Relationship Among Water Flow Units

The lines that connect the boxes and the numbers associated with them can be thought of as bridges that relate two units directly and all other units indirectly. Multiply when bringing to a larger box and divide when converting to a smaller box. To convert from one unit to another one can use as many bridges as necessary.

Dimensional Analysis

In the previous section the term "dimensional analysis" was introduced when it was demonstrated how to divide and multiply units as well as numbers. This is important in all calculations, not only in plant computations but also in laboratory work.

The fact that a tank holds 5 000 gal of water conveys a meaning similar to the fact that a beaker contains 250 ml of water. The 5 000 in the first case and the 250 in the second are the numerical values whereas gallons and milliliters (ml) are the dimensions. Heretofore, these dimensions have been called units. In the examples in the section on rates, the numerical values and dimensions (units) were considered at the same time. But to work out conversion factors, that is, change one unit or dimension to another, one need not involve numbers at all. When this is done, that is, treating dimensions mathematically, one has what is called "dimensional analysis."

To convert fps to mph, one needs to convert seconds to hours and feet to miles.

$$\frac{\text{ft}}{\text{sec}} \times \frac{\text{sec}}{\text{hr}} \times \frac{\text{mi}}{\text{ft}} = \frac{\text{mi}}{\text{hr}}$$

One must now determine a numerical value of the constant relating to the two dimensions.

$$1.00\frac{\text{ft}}{\text{s}} = 1.00\frac{\text{ft}}{\text{s}} \times \frac{3\,600\text{ s}}{\text{hr}} \times \frac{1\text{ mi}}{5\,280\text{ ft}} = \frac{3\,600\text{ mi}}{5\,280\text{ hr}} = 0.6818\,\frac{\text{mi}}{\text{hr}} = 0.6218\text{ mph}$$

The preceding procedure is useful for converting from English to metric units. See Table IV-3 for common conversions.

It is supposed that the water is clarified best by the use of 10 mg/l of alum. How many pounds of alum must be fed for 5 mil gal of water?

The first problem is to convert mg/l to lb/gal.

$$\frac{\text{mg}}{\text{l}} \times \frac{\text{l}}{\text{gal}} \times \frac{\text{lb}}{\text{mg}} = \frac{\text{lb}}{\text{gal}}$$

Next, insert numerical values:

$$1.00\text{ mg/l} = \frac{\text{mg}}{\text{l}} \times 3\,785\frac{\text{l}}{\text{gal}} \times \frac{\text{lb}}{454 \times 10^3\text{mg}} = \frac{3.785\text{ lb}}{0.454 \times 10^6\text{gal}} = 8.34 \times 10^{-6}\text{ lb/gal}$$

Because 10 mg/l of alum was used for 5 mil gal of water, $10 \times 8.34 \times 10^6$ lb/gal $\times\ 5 \times 10^6$ gal = 417 lb alum is necessary to clarify 5 mil gal of water.

Averages and Means

An average is a simple number used to represent a group of measurements or data. Averages are of different kinds. The commonly used average is the arithmetic mean. It is what is usually meant by the "mean" or "average." If X is the value of the individual measurements and N is their number, then

$$\text{arithmetic mean} = \overline{X} = \frac{X_1 + X_2 + X_3 + \ldots + X_N}{N}$$

If X_1 appears f_1 times and X_2 appears f_2 times, etc., then

$$\overline{X} = \frac{X_1f_1 + X_2f_2 + X_3f_3 + \ldots + X_Nf_N}{N}$$

where N = sum of all f's.

Weighted averages

The previous equation shows that each X must be weighted according to the number of times that it occurs. In Table IV-4 the rate of pumping is constant beginning on the hour and continuing until the next recorded hour.

TABLE IV-4

Recorded Rate of Pumping

Hour	mil gal/hr
12:00 Midnight	2.0
1:00 AM	2.5
3:00 AM	2.7
6:00 AM	3.0
7:00 AM	7.0
12:00 Noon	6.0
2:00 PM	4.0
5:00 PM	3.0
9:00 PM	2.0

The desired hourly average flow in million gallons over the 24-hr period is to be computed. Merely to add the values in the table and divide by 9, one would produce 3.58. This would not be correct because the rates given in the table do not hold for equal times. Each rate must be weighed according to the hours run at that rate.

$$
\begin{aligned}
2.5 \times 2 &= 5.0 \\
2.7 \times 3 &= 8.1 \\
3 \times 1 &= 3.0 \\
7 \times 5 &= 35.0 \\
6 \times 2 &= 12.0 \\
4 \times 3 &= 12.0 \\
3 \times 4 &= 12.0 \\
2 \times 3 &= 6.0 \\
\text{Sum} &= 95.1
\end{aligned}
$$

$$\text{Weighted average} = \frac{95.1}{24} = 3.96 \text{ mil gal/hr for the correct average.}$$

The median

Another type of average is the median. This is determined by arranging the values in descending order and selecting the value halfway down. Another method of obtaining the mean is to cross out the largest and smallest values, continuing until only one is left. If two are left, the average of these two is taken. Half the values are larger than the median and half are smaller. It has the advantage of not being affected by an abnormally large or too small value of a single quantity.

The mode

Another representative value is the mode. The mode is that value which in any series of observations occurs more often that any other. Its chief use in sanitation studies is in connection with the MPN (most probable number) values for coliform density. In general operational practice it is not customary to use the mode. Arithmetic averages are commonly used.

Algebraic Signs and Symbols

An algebraic symbol is a letter representing a definite operation or relation.

Addition a+b
Subtraction a-b
Multiplication axb or a·b or ab
Division a/b or a÷b or $\frac{a}{b}$
Powers $a \times a = a^2$
$a \times a \times a \times \ldots n \text{ times} = a^n$
Square root $\sqrt{a} = a^{1/2}$
Fractional exponent $a^{m/n} = (a^{1/n})^m = (a^m)^{1/n} = \sqrt[n]{a^m}$

Meaning of signs

The signs + and - have two distinct meanings in algebra. They are signs of operation for addition and subtraction. For addition or subtraction, they are read "plus" and "minus." They are also signs of quality and are read "positive" and "negative." Positive and negative are "opposites." One must be given a meaning, whereupon the other becomes the opposite. If distances upward are arbitrarily marked positive, distances downward are negative. Credit may be marked positive and indebtedness negative. Temperatures above zero are positive and below zero are negative.

Examples of algebraic operations are

$$(+a) + (+b) = a + b$$
$$(-a) + (+b) = -a + b = b - a$$
$$(-a) + (-b) = -a - b = -(a + b)$$
$$(-a)(+b) = -ab$$
$$(+a)(+b) = ab$$
$$(-a)(-b) = +ab$$
$$n(a + b) = na + nb$$
$$-n(a + b) = -na - nb$$
$$-n(a - b) = -na + nb = nb - na$$
$$\frac{-a}{+b} = -\frac{a}{b} \qquad \frac{-a}{-b} = \frac{a}{b}$$

Signs of aggregation

The signs of aggregation that are usually used are parentheses ( ), brackets [ ], braces { }, and vinculum $\overline{}$. An expression within a pair of signs of aggregation must be treated as a unit. To remove signs of aggregation, begin with the innermost pair and work outward.

$$6 - [2 - (8 - 2)] = 10$$

Rules for Exponents

$$a^m \times a^n = a^{m+n}$$
$$\frac{a^m}{a^n} = a^{m-n}$$
$$(a^m)^n = a^{mn}$$

$$(a^m)^{1/n} = a^{m/n}$$

$$a^{-m} = \frac{1}{a^m}$$

Solving for the Unknown

In many calculations, the unknown X is part of a multiplication or division calculation.

When solving for the unknown X in a multiplication or division problem, two basic objectives will aid one in his calculations:

1. X must always be in the numerator.
2. All other numbers must be cleared from the "X side" of the equation.

To accomplish these objectives, one may move any of the numbers to either side of the equation providing he follows one basic rule:

Whatever is done to one side of the equation must be done to the other side of the equation. A person may therefore add or subtract equal quantities from both sides of an equation. Or both sides of an equation can be multiplied or divided by the same quantity.

Example 1

Solve for the feet per minute velocity. Based on the concept of volume, the mathematical setup is

$$(4\text{ ft})\ (1.5\text{ ft})\ (X\text{ fpm}) = 1\ 100\text{ cfm}$$

When solving for the unknown X, two objectives must be met

1. X must be in the numerator.
2. All other numbers must be cleared from the "X side" of the equation.

In this example, the first objective is satisfied. Once X is in the numerator, this becomes the "X side" of the equation. All other numbers (4 and 1.5) must be cleared from the "X side." If no denominators are shown, 1 is assumed to be the denominator of the fraction:

$$\frac{(4)\ (1.5)\ (X)}{1} = \frac{1\ 100}{1}$$

The 4 and 1.5 must be moved to the other side of the equation. This can be done by dividing both sides by $4 \cdot 1.5$.

$$X\text{ fpm} = \frac{1\ 100}{(4)(1.5)}$$

$$X = 183.3\text{ fpm}$$

The unknown X has now been solved.

Example 2

Given the following equation, solve for the unknown X.

$$730 = \frac{X}{3\ 847}$$

X is already in the numerator; one must merely clear for X by multiplying both sides by 3.847:

$$(3\ 847)\ (730) = X$$

Example 3

Given the following equation, solve for the unknown X.

$$689 = \frac{5\,400\,000}{X}$$

X must be in the numerator. It can be moved from one side of the equation to the other by multiplying both sides of the equation by X and cancelling the Xs on the right.

$$(X)\,(689) = 5\,400\,000$$

Clearing all numbers from the "X side" of the equation,

$$X = \frac{5\,400\,000}{689}$$

Example 1

Given the equation below, solve for the unknown X.

$$7.2 = X + 0.2$$

Clear for the unknown X by subtracting 0.2 from both sides of the equation.

$$7.2 - 0.2 = X$$
$$7.0 = X$$

Example 2

Given the equation below, solve for the unknown X.

$$7.5 - X = 1.1$$

To solve for positive X, multiply both sides of the equation by -1

$$-7.5 + X = -1.1$$

Then add 7.5 to both sides of the equation

$$X = -1.1 + 7.5$$
$$X = 6.4$$

Temperature

The formulas normally used for converting Fahrenheit (F) to Celsius (C, previously called centigrade) are

Fahrenheit to Celsius	*Celsius to Fahrenheit*
$C = 5/9\,(F - 32)$	$F = 9/5\,(C) + 32$

Because these formulas have parentheses around different terms, and one requires the addition of 32 while the other requires the subtraction of 32, they can be difficult conversions to remember if not used constantly.

Another method of converting these two terms is more easily remembered and will remain the same regardless of whether converting from Fahrenheit to Celsius or vice versa. This method will consist of three steps:

1. Add 40
2. Multiply by 5/9 or 9/5
3. Subtract 40

Step 2 involves the only variable in these conversions. The decision of whether to multiply by 5/9 or 9/5 is dependent upon whether moving from a larger number to a smaller number or vice versa—5/9 being just over half and 9/5 being almost 2.

Example 1

The influent to a treatment plant has a temperature of 75F. What is this temperature expressed in degrees Celsius?

Step 1: (add 40)

$$\begin{array}{r} 75 \\ +40 \\ \hline 115 \end{array}$$

Step 2: (Multiply by 5/9 or 9/5)

This example converts Fahrenheit to Celsius. One should expect a *smaller* number because the boiling point of water (at sea level) is 212F and 100C, and thus, *multiply by 5/9* (rounding answer).

$$(5/9)(115/1) = \frac{575}{9} = 64$$

Step 3: (Subtract 40)

$$\begin{array}{r} 64 \\ -40 \\ \hline 24 \end{array}$$

75F = 24C

Example 2:

The effluent of a treatment plant is 23C. What is this expressed in degrees Fahrenheit?

Step 1: (add 40)

$$\begin{array}{r} 23 \\ +40 \\ \hline 63 \end{array}$$

Step 2: (Multiply by 5/9 or 9/5) Celsius to Farenheit. One should expect a larger number and thus, multiply by 9/5.

$$(9/5)(63/1) = \frac{567}{5} = 113$$

Step 3: (Subtract 40)

$$\begin{array}{r} 113 \\ -\ 40 \\ \hline 73 \end{array}$$

23 C = 73 F

Chapter V
Applied Hydraulics

Introduction

Hydraulics is the study of liquid in motion or under pressure. In this handbook the study of hydraulics will be confined to the behavior of water, with particular reference to public water-supply systems. An understanding of hydraulics is necessary for the proper operation of a water-distribution system.

Pressure

Water flows in a water system when it is under a force that makes it move. The force on a unit area of water is termed *pressure.* The pressure in a water system is a measure of the height to which water theoretically will rise in an imaginary sys- standpipe open to atmospheric pressure. Pressure can be *static;* i.e., it exists although the water does not flow. Or pressure can be *dynamic,* existing as moving "energy."

All objects have weight because they are acted on by gravity. When a l-lb brick is placed on a table of unit (e.g., 1 sq in.) area, it exerts a force of 1 lb on the table and the pressure on the table is 1 lb/sq in. (1 psi). Two stacked bricks on the same table exert a 2-lb force and a 2-psi pressure. If the area of contact between the table and the brick is doubled the pressure is halved; tripled, and the pressure is reduced to ⅓. The area of contact and not the weight alone influences the pressure.

One can consider a body of water as being made up of many "bricks." A "brick" 1 in. on each end weighs 0.036 lb and adds its weight to the bricks of water beneath it. A column of water 1 ft (12 in.) high therefore exerts a total weight of the column of 0.43 lb. The pressure is 0.43 lb/sq in. (psi). The pressure of the force per unit area at the bottom of a water column depends directly, and only, on the height of the column. Thus a 1 sq ft column, 1 ft high with a total weight of 62.4 lb exerts a hydrostatic pressure of 0.43 psi. A 10-ft column impresses 4.3 psi. (See Fig. V-1.)

If the water in the column is permitted to empty horizontally from the bottom of the column, then the water will begin to flow under the hydrostatic pressure applied by the height of the column. The flowing water will have little hydrostatic pressure, but it will have gained moving, dynamic pressure or kinetic energy. The hydrostatic pressure is static potential energy that is converted into moving (kinetic) energy.

One can add energy to a water system and thereby increase hydrostatic and/or

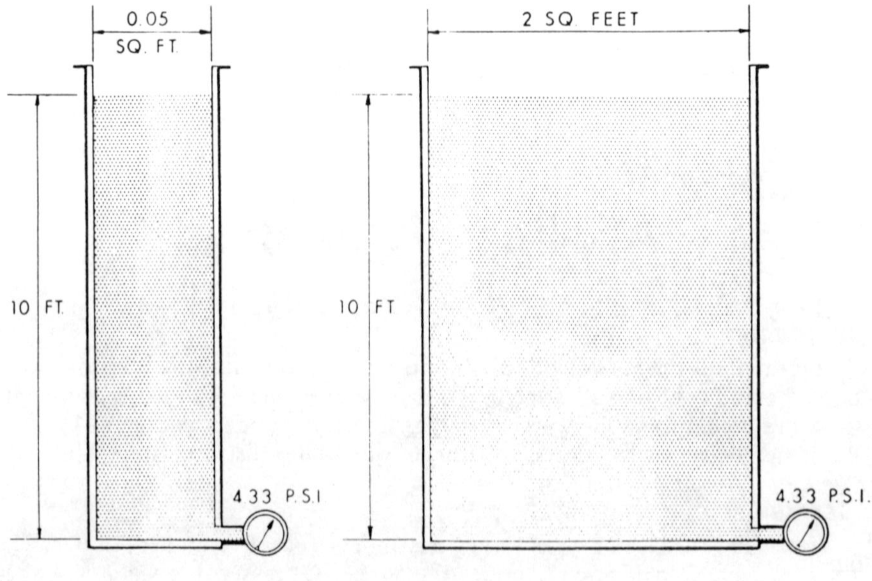

Fig. V-1. Hydraulic Head Depends Only on Column Height

dynamic pressure. This is a function of a pump. Thus water is pumped into elevated storage. The hydrostatic pressure (height) to which the water can be pumped is equivalent to pressure, less losses, at the output of the pump.

Pressure is usually measured in pounds per square inch, or in head (feet of height of the water column). The pressure in pounds per square inch is equal to 0.433 times feet of head. It can be measured with pressure gages, some of which read in pounds per square inch and some of which read in feet of head.

Velocity

The speed at which water moves is called velocity V. Velocity is usually measured in either miles per hour (mph) or feet per second (fps). A rapidly flowing stream, for example, might move at a velocity of 5 mph or 7.3 fps. The velocity of 1 mph is equal to 1.46 fps; alternatively 1 fps is equal to 0.683 mph.

The quantity of water Q that flows through a pipe or a ditch depends upon velocity V and cross-sectional area A of the flow at right angles to its direction. Stated mathematically, $Q = AV$. For example, a square flume or ditch 1 ft wide and 1 ft deep with water would have an area A at right angles to the direction of flow of 1 sq ft. If the water were flowing in this ditch with a velocity V of 1 fps, the quantity Q flowing would be 1 cu ft/s (cfs). The 1 cfs is equal to 7.48 gal/s since there are 7.48 gal in 1 cu ft. The 1 cfs would also be equal to 60 cu ft/min (cfm), or multiplying by 7.48 gal/cu ft, 449 gal/min (gpm). Because there are 1440 min in a day, one can multiply the gallons per minute by 1440 and obtain the total gallons that flow in a 24-hr day. One cfs, or 449 gpm, multiplied by 1440 is equal to 646 000 gal/day (gpd).

This leads to the obvious and important relation that the quantity of water flowing is equal to the cross sectional area of flow in the pipe times the velocity. When the quantity of flow is calculated in cubic feet per second it can then be changed to gallons per minute by multiplying by 449, as previously explained, or to gallons per day by multiplying by 646 000.

Friction Loss

Although water will always flow downhill and appears to do so without anything tending to resist or slow its flow, there is friction due to an apparent roughness between the flowing water and the surface of the pipe in which the water flows.

Theoretically, the flow in a pipe under pressure varies as the square root of the hydraulic gradient or loss of head, other conditions being equal. To state this idea in another way, the loss of head or friction loss caused by flow in the pipe varies as the square of the quantity of flow.

Friction losses in a pipeline depend upon the velocity or rate of flow, the size of the pipe (that is, its diameter), its length, and the apparent roughness of the surface in contact with the water. The degree of pipe roughness is commonly denoted by a *C* factor, which is a coefficient in the Hazen-Williams formula for determining flow. Conversely flow rate (quantity) varies with friction. For most pipe materials *C* ranges from 90 to 140. High *C* values imply less friction

Head or friction losses for 100-ft lengths of pipe of varying sizes for a *C* factor of 100 are tabulated in Table V-1. For example, the head loss that would occur through 100 ft of 2-in. pipe for a flow of 100 gpm is ascertained by following the "US gpm" column to 100, then moving across the "Loss in ft" column for 2-in. pipe. The head loss is read as 35.8 ft. Multiplying this figure by 0.433 yields a 15.5 psi loss.

Thus if the pressure in the main where a 2-in. tap is made is 50 psi, the pressure 100 ft away at a point where a service might enter a house, for example, would only be 34.5 psi (50 minus 15.5) when the flow through the service is 100 gpm. The friction loss would be twice as much for 200 ft of pipe and 10 times as much for 1000 ft of pipe if the average flow rate were maintained at 100 gpm. (The friction loss for lengths of pipe other than100 ft is determined by multiplying the values in the table by the number of 100-ft lengths of pipe through which the water actually flows.)

When a pipe has a *C* factor other than 100, it is possible to adjust back to a *C* of 100 by multiplying the flow by 100/*C* and then entering the table with the recomputed flow value. Thus for the previous example, had the *C* factor been 125 instead of 100, the adjusted flow rate would have been 100 x 100/125 = 80 gpm. Entering the table at 80 gpm and crossing over to the 2-in.-pipe column yields a loss of 23.2 ft. Thus, the smoother pipe "saves" 12.1 (35.8 - 23.2) ft of head for each 100 ft of pipe used.

In addition to friction losses attributable to flow through straight sections of pipe, losses occur through fittings and orifices. The effect on flow of some types of fittings is shown in Fig. V-2.

Hydraulic Gradient

To determine flow of water in pipelines, the idea of hydraulic gradient is help-

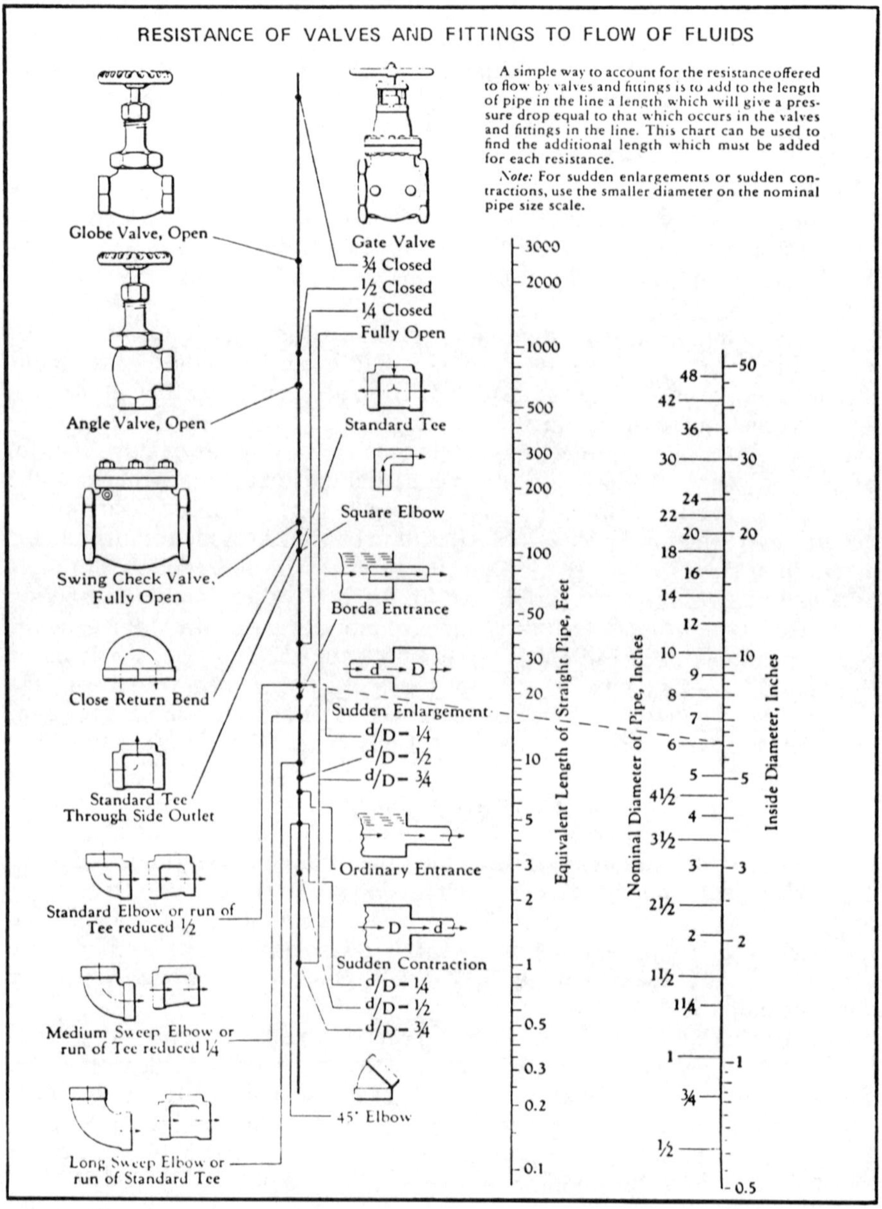

Permission to Reprint Courtesy
Crane Company, 1974

Fig. V-2. Fittings Increase Pressure Losses

ful. The hydraulic gradient is nothing more than the slope of the imaginary line connecting the levels to which water would rise in freely vented standpipes were they constructed along a pipeline. If the pipe is uniform in size and there are no branch lines or takeoffs along the line, then the hydraulic grade line would be a straight line. The slope of this line determines the quantity of flow. A typical example is shown in Fig. V-3 of a pipeline of different pipe sizes with a particular gradient along each section of pipe of one size. The gradient becomes steeper as the pipe becomes smaller. The slope or the steepness of the hydraulic grade line, which represents the elevation to which water would rise in a standpipe, is the head loss as given in the Friction Loss Table for the particular pipe plus the difference in elevation of the two ends of the pipe divided by the horizontal distance.

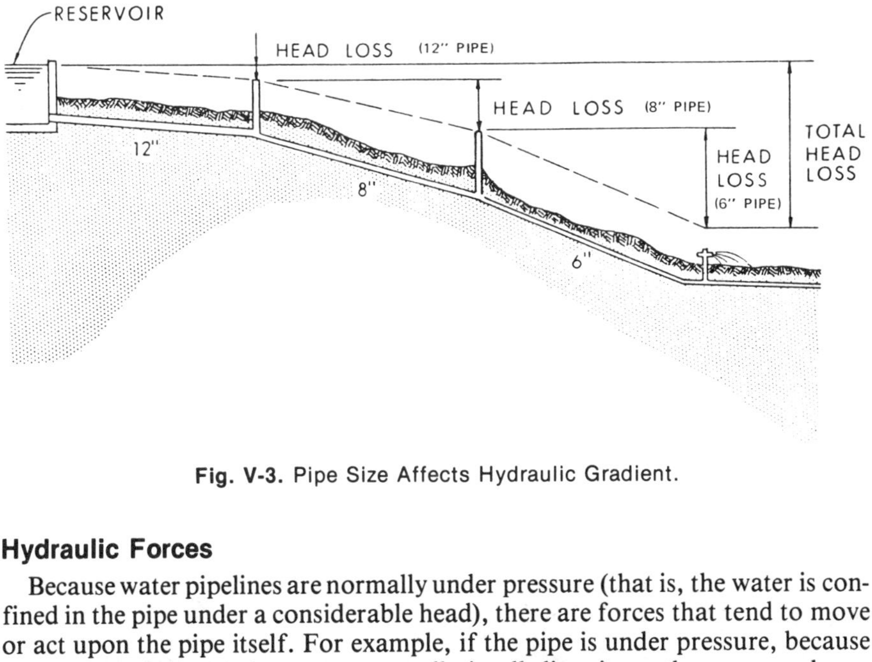

Fig. V-3. Pipe Size Affects Hydraulic Gradient.

Hydraulic Forces

Because water pipelines are normally under pressure (that is, the water is confined in the pipe under a considerable head), there are forces that tend to move or act upon the pipe itself. For example, if the pipe is under pressure, because water exerts hydrostatic pressure equally in all directions, the pressure has a tendency to stretch or to expand the pipe. Consequently, the pipe must be strong enough to resist internal pressure or the tendency to expand.

When there is a closed valve at the end of a line under pressure, the pressure of the water is applied to the valve. Unless the valve is blocked, shackled, or fastened to the pipeline with flange bolts, it can blow off. Similarly, if a riser is placed in the line and a cap is placed on the end of the riser, the cap will blow off unless it is strong enough to resist the pressure of the water against it.

Wherever there is a change of direction in a pipeline, there is an unbalanced force that tends to move the pipe. With a 90-deg bend in a pipeline, there will be pressure against the bend that will tend to blow the bend off the remainder of the

TABLE V-1

FRICTION LOSS OF WATER PER 100 FEET LENGTH OF PIPE (INSIDE DIAMETER)
Based on Williams & Hazen Formula Using Constant 100
Sizes of Standard Pipe in Inches

½" Pipe		3/4" Pipe		1" Pipe		1¼" Pipe		1½" Pipe		2" Pipe		2½" Pipe		3" Pipe		4" Pipe		5" Pipe		6" Pipe		U.S. Gals. Per min.
Vel. 'per Sec.	Loss in Feet	Vel. 'per Sec.	Loss in Feet	Vel. 'per Sec.	Loss in Feet	Vel. 'per Sec.	Loss in Feet	Vel. 'per Sec.	Loss in Feet	Vel. 'per Sec.	Loss in Feet	Vel. 'per Sec.	Loss in Feet	Vel. 'per Sec.	Loss in Feet	Vel. 'per Sec.	Loss in Feet	Vel. 'per Sec.	Loss in Feet	Vel. 'per Sec.	Loss in Feet	
2.10	7.4	1.20	1.9																			2
4.21	27.0	2.41	7.0	1.49	2.14	.86	.57	.63	.26													4
6.31	57.0	3.61	14.7	2.23	4.55	1.29	1.20	.94	.56	.61	.20											6
8.42	98.0	4.81	25.0	2.98	7.8	1.72	2.03	1.26	.95	.82	.33	.52	.11									8
10.52	147.0	6.02	38.0	3.72	11.7	2.14	3.05	1.57	1.43	1.02	.50	.65	.17	.45	.07							10
		7.22	53.0	4.46	16.4	2.57	4.3	1.89	2.01	1.23	.79	.78	.23	.54	.10							12
		9.02	80.0	5.60	25.0	3.21	6.5	2.36	3.00	1.53	1.08	.98	.36	.68	.15							15
		10.84	108.2	6.69	35.0	3.86	9.1	2.83	4.24	1.84	1.49	1.18	.50	.82	.21							18
		12.03	136.0	7.44	42.0	4.29	11.1	3.15	5.20	2.04	1.82	1.31	.61	.91	.25	.51	.06					20
				9.30	64.0	5.36	16.6	3.80	7.30	2.55	2.73	1.63	.92	1.13	.38	.64	.09					25
				11.15	89.0	6.43	23.0	4.72	11.0	3.06	3.84	1.96	1.29	1.36	.54	.77	.13	.49	.04			30
				13.02	119.0	7.51	31.2	5.51	14.7	3.57	5.10	2.29	1.72	1.59	.71	.89	.17	.57	.06			35
				14.88	152.0	8.58	40.0	6.30	18.8	4.08	6.6	2.61	2.20	1.82	.91	1.02	.22	.65	.08			40
						9.65	50.0	7.08	23.2	4.60	8.2	2.94	2.80	2.04	1.15	1.15	.28	.73	.09			45
						10.72	60.0	7.87	28.4	5.11	9.9	3.27	3.32	2.27	1.38	1.28	.34	.82	.11	.57	.04	50
						11.78	72.0	8.66	34.0	5.62	11.8	3.59	4.01	2.45	1.58	1.41	.41	.90	.14	.62	.05	55
						12.87	85.0	9.44	39.6	6.13	13.9	3.92	4.65	2.72	1.92	1.53	.47	.98	.16	.68	.06	60
						13.92	99.7	10.23	45.9	6.64	16.1	4.24	5.4	2.89	2.16	1.66	.53	1.06	.19	.74	.076	65
						15.01	113.0	11.02	53.0	7.15	18.4	4.58	6.2	3.18	2.57	1.79	.63	1.14	.21	.79	.08	70
						16.06	129.0	11.80	60.0	7.66	20.9	4.91	7.1	3.33	3.00	1.91	.73	1.22	.24	.85	.10	75
						17.16	145.0	12.59	68.0	8.17	23.7	5.23	7.9	3.63	3.28	2.04	.81	1.31	.27	.91	.11	80
						18.21	163.8	13.38	75.0	8.68	26.6	5.56	8.1	3.78	3.54	2.17	.91	1.39	.31	.96	.12	85
						19.30	180.0	14.71	84.0	9.19	29.4	5.88	9.8	4.09	4.06	2.30	1.00	1.47	.34	1.02	.14	90
								14.95	93.0	9.70	32.6	6.21	10.8	4.22	4.33	2.42	1.12	1.55	.38	1.08	.15	95
8" PIPE								15.74	102.0	10.21	35.8	6.54	12.0	4.54	4.96	2.55	1.22	1.63	.41	1.13	.17	100
								17.31	122.0	11.23	42.9	7.18	14.5	5.00	6.0	2.81	1.46	1.79	.49	1.25	.21	110
								18.89	143.0	12.25	50.0	7.84	16.8	5.45	7.0	3.06	1.17	1.96	.58	1.36	.24	120
								20.46	166.0	13.28	58.0	8.48	18.7	5.91	8.1	3.31	1.97	2.12	.67	1.47	.27	130
.90	.08							22.04	190.0	14.30	67.0	9.15	22.3	6.35	9.2	3.57	2.28	2.29	.76	1.59	.32	140
.96	.09									15.32	76.0	9.81	25.5	6.82	10.5	3.82	2.62	2.45	.88	1.70	.36	150
1.02	.10									16.34	86.0	10.46	29.0	7.26	11.8	4.08	2.91	2.61	.98	1.82	.40	160
1.08	.11	10" PIPE								17.36	96.0	11.11	34.1	7.71	13.3	4.33	3.26	2.77	1.08	1.92	.45	170
1.15	.13									18.38	107.0	11.76	35.7	8.17	14.0	4.60	3.61	2.94	1.22	2.04	.50	180
1.21	.14									19.40	116.0	12.42	39.6	8.63	15.5	4.84	4.01	3.10	1.35	2.16	.55	190
1.28	.15									20.42	129.0	13.07	43.1	9.08	17.8	5.11	4.4	3.27	1.48	2.27	.62	200
1.40	.18	.90	.06							22.47	154.0	14.38	52.0	9.99	21.3	5.62	5.2	3.59	1.77	2.50	.73	220

1.53	.22	.98	.07							24.51	182.0	15.69	61.0	10.89	25.1	6.13	6.2	3.92	2.08	2.72	.87			240
1.66	.25	1.06	.08							26.55	211.0	16.99	70.0	11.80	29.1	6.64	7.2	4.25	2.41	2.95	1.00			260
1.79	.28	1.15	.09									18.30	81.0	12.71	33.4	7.15	8.2	4.58	2.77	3.18	1.14			280
1.91	.32	1.22	.11									19.61	92.0	13.62	38.0	7.66	9.3	4.90	3.14	3.40	1.32			300
2.05	.37	1.31	.12	12" PIPE								20.92	103.0	14.52	42.8	8.17	10.5	5.23	3.54	3.64	1.47			320
2.18	.41	1.39	.14									22.22	116.0	15.43	47.9	8.68	11.7	5.54	3.97	3.84	1.62			340
2.30	.45	1.47	.15			14" PIPE						23.53	128.0	16.34	53.0	9.19	13.1	5.87	4.41	4.08	1.83			360
2.43	.50	1.55	.17	1.08	.069							24.84	142.0	17.25	59.0	9.69	14.0	6.19	4.86	4.31	2.00			380
2.60	.54	1.63	.19	1.14	.075							26.14	156.0	18.16	65.0	10.21	16.0	6.54	5.4	4.55	2.20			400
2.92	.68	1.84	.23	1.28	.95									20.40	78.0	11.49	19.8	7.35	6.7	5.11	2.74			450
3.19	.82	2.04	.28	1.42	.113	1.04	.06							22.70	98.0	12.77	24.0	8.17	8.1	5.68	2.90			500
3.52	.97	2.24	.33	1.56	.135	1.15	.07							24.96	117.0	14.04	28.7	8.99	9.6	6.25	3.96			550
3.84	1.14	2.45	.39	1.70	.159	1.25	.08							27.23	137.0	15.32	33.7	9.80	11.3	6.81	4.65			600
4.16	1.34	2.65	.45	1.84	.19	1.37	.09									16.59	39.0	10.62	13.2	7.38	5.40			650
4.46	1.54	2.86	.52	1.99	.22	1.46	.10									17.87	44.9	11.44	15.1	7.95	6.21			700
4.80	1.74	3.06	.59	2.13	.24	1.58	.11	16" PIPE								19.15	51.0	12.26	17.2	8.50	7.12			750
5.10	1.90	3.26	.66	2.27	.27	1.67	.13									20.42	57.0	13.07	19.4	9.08	7.96			800
5.48	2.20	3.47	.75	2.41	.31	1.79	.14	1.36	.08							21.70	64.0	13.89	21.7	9.65	8.95			850
5.75	2.46	3.67	.83	2.56	.34	1.88	.16	1.44	.084	20" PIPE						22.98	71.0	14.71	24.0	10.20	10.11			900
6.06	2.87	3.88	.91	2.70	.38	2.00	.18	1.52	.095									15.52	26.7	10.77	11.20			950
6.38	2.97	4.08	1.03	2.84	.41	2.10	.19	1.60	.10	1.02	.04							16.34	29.2	11.34	12.04			1000
7.03	3.52	4.49	1.19	3.13	.49	2.31	.23	1.76	.12	1.12	.04							17.97	34.9	12.48	14.55			1100
7.66	4.17	4.90	1.40	3.41	.58	2.52	.27	1.92	.14	1.23	.05							19.61	40.9	13.61	17.10			1200
8.30	4.85	5.31	1.62	3.69	.67	2.71	.32	2.08	.17	1.33	.06									14.72	18.4			1300
8.95	5.50	5.71	1.87	3.98	.78	2.92	.36	2.24	.19	1.43	.06	24" PIPE								15.90	22.60			1400
9.58	6.24	6.12	2.13	4.26	.89	3.15	.41	2.39	.21	1.53	.07									17.02	25.60			1500
10.21	7.00	6.53	2.39	4.55	.98	3.34	.47	2.56	.24	1.63	.08			30" PIPE						18.10	26.9			1600
11.50	8.78	7.35	2.95	5.11	1.21	3.75	.58	2.87	.30	1.84	.10	1.28	.04											1800
12.78	10.71	8.16	3.59	5.68	1.49	4.17	.71	3.19	.37	2.04	.12	1.42	.05											2000
14.05	12.78	8.98	4.24	6.25	1.81	4.59	.84	3.51	.44	2.25	.15	1.56	.06											2200
15.32	14.2	9.80	5.04	6.31	2.08	5.00	.99	3.83	.52	2.45	.17	1.70	.07	1.09	.02									2400
		10.61	5.81	7.38	2.43	5.47	1.17	4.15	.60	2.66	.20	1.84	.08	1.16	.027									2600
		11.41	6.70	7.95	2.75	5.84	1.32	4.47	.68	2.86	.23	1.98	.09	1.27	.03									2800
		12.24	7.62	8.52	3.15	6.01	1.49	4.79	.78	3.08	.27	2.13	.10	1.37	.037									3000
		13.05	7.8	9.10	3.51	6.68	1.67	5.12	.88	3.27	.30	2.26	.12	1.46	.041									3200
		14.30	10.08	9.95	4.16	7.30	1.97	5.59	1.04	3.59	.35	2.49	.14	1.56	.047									3500
		15.51	13.4	10.80	4.90	7.98	2.36	6.07	1.20	3.88	.41	2.69	.17	1.73	.05									3800
				11.92	5.88	8.76	2.77	6.70	1.44	4.29	.49	2.99	.20	1.91	.07									4200
				12.78	6.90	9.45	3.22	7.18	1.64	4.60	.56	3.20	.22	2.04	.08									4500
				14.20	8.40	10.50	3.92	8.01	2.03	5.13	.68	3.54	.27	2.26	.09									5000
						11.55	4.65	8.78	2.39	5.64	.82	3.90	.33	2.50	.11									5500
						12.60	5.50	9.58	2.79	6.13	.94	4.25	.38	2.73	.13									6000
						13.65	6.45	10.39	3.32	6.64	1.10	4.61	.45	2.95	.15									6500
						14.60	7.08	11.18	3.70	7.15	1.25	4.97	.52	3.18	.17									7000
								12.78	4.74	8.17	1.61	5.68	.66	3.64	.23									8000
								14.37	5.90	9.20	2.01	6.35	.81	4.08	.28									9000
								15.96	7.19	10.20	2.44	7.07	.98	4.54	.33									10000
										12.25	3.41	8.50	1.40	5.46	.48									12000
										14.30	4.54	9.95	1.87	6.37	.63									14000
												11.38	2.40	7.28	.81									16000
												12.76	2.97	8.18	1.02									18000
												14.20	3.60	9.10	1.23									20000

Thrust at Fittings Due to Water Pressure

Size *in*	Pressure *psi*	Tees Dead Ends	90-deg bend	45-deg bend	22.5-deg bend
3	–100	1220	1720	931	474
	150	1330	1880	1020	519
	200	1610	2270	1230	628
4	–100	1910	2750	1490	759
	150	2020	2860	1550	790
	200	2350	3330	1800	919
6	–100	3830	5420	2930	1500
	150	4100	5800	3140	1600
	200	4750	6710	3630	1850
8	–100	6580	9310	5040	2570
	150	6980	9870	5340	2720
	200	7910	11200	6050	3090
10	– 50	9820	13900	7510	3830
	100	10300	14600	7890	4020
	150	11300	15900	8620	4400
	200	12300	17400	9450	4810
12	– 50	13800	19500	10600	5390
	100	14600	20600	11200	5700
	150	15900	22500	12200	6210
	200	17200	24400	13200	6720
14	– 50	18600	26300	14300	7270
	100	19800	28000	15200	7740
	150	21600	30600	16500	8440
	200	23300	32900	17800	9080
16	– 50	24000	33900	18400	9370
	100	25700	36300	19700	10000
	150	27800	39400	21300	10900
	200	30300	42800	23200	11800
18	– 50	30000	42400	23000	11700
	100	32300	45600	24700	12600
	150	35000	49500	26800	13700
	200	38300	54100	29300	14900
20	– 50	36700	51900	28100	14300
	100	39700	56100	30300	15500
	150	43000	60800	32900	16800
	200	47400	67000	36300	18500
24	– 50	52300	74000	40000	20400
	100	56800	80300	43500	22200
	150	61700	87300	47200	24100
	200	68200	96400	52200	26600
30	– 50	82400	117000	63100	32200
	100	88200	125000	67500	34400
	150	96700	137000	74000	37700
	200	106000	150000	81100	41400
36	– 50	118000	167000	90400	46100
	100	127000	179000	96900	49400
	150	139000	197000	107000	54300
	200	152000	215000	116000	59300

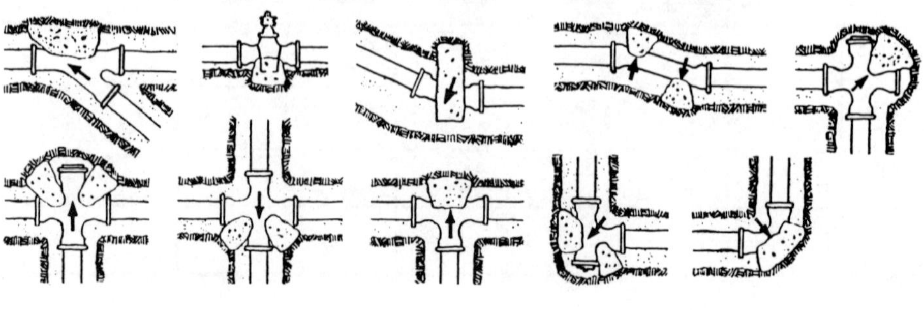

Fig. V-4. Flow Direction Changes Induce Forces on Pipe.

pipe. This pressure will act upon the area of the pipe diameter and in the direction in which the bent surface of the pipe faces. The actual calculation of the force against a 90-deg bend, or against any other bend, is a somewhat complicated problem in trigonometry involving a resolution of forces acting at angles. It should be realized that a change in the vertical or horizontal direction will cause a thrust on the pipe (see the table that is part of Fig. V-4). Therefore, all bends must be anchored.

The usual method for anchoring bends in which the unbalanced thrust is upward is to fasten sufficient concrete to the pipeline to counteract the uplift force. In the case of horizontal bends, the pipe is normally blocked against undisturbed earth so that it cannot move. An examination of the amount of the force will give adequate indication of the need for careful anchorage of pipelines at bends. Anyone who has worked with water pipelines for very long has experienced this force when an improperly anchored bend has moved under pressure. The amount of concrete required to anchor a horizontal bend depends on the strength of the soil. The methods of placing concrete to keep the joint accessible are shown in Fig. V-4. The area, in square feet, of concrete that must bear against the side of the trench is determined by dividing the thrust, in pounds, by the safe bearing load of the soils. Safe bearing loads of soil for horizontal thrusts when the depth of cover over the pipe exceed 2 ft are shown in Table V-2.

TABLE V-3
Safe Bearing Loads

Soil	Safe Bearing Load—*psf*
Muck or peat*	0
Soft clay	1 000
Sand	2 000
Sand and gravel	3 000
Sand and gravel cemented with clay	4 000
Hard shale	10 000

*In muck or peat, all thrusts are resisted by pipes or tie rods to solid foundations or by removal of muck or peat and replacement with ballast of sufficient stability to resist thrusts.

In addition to the normal pressure exerting a force, a valve closed too rapidly causes the water to come to a sudden stop, thus causing waves of high pressure that oscillate back and forth in the isolated pipe section. This is called "water hammer." Where this is likely to occur, additional blocking will be required. For more information on thrust blocks see Chapter VIII.

Chapter VI

Reservoirs

Introduction

Chapter II briefly covers some aspects of operational storage. This chapter is concerned with distribution reservoirs — specifically, reservoirs that serve as operational storage of treated water. Reservoirs are a necessary part of any distribution system if the water utility is to meet peak demands and do so at a minimum cost.

Types of Reservoirs

Reservoirs may be of several different types. Often a water utility will have more than one type of reservoir. Early reservoirs were of the cut-and-fill earth embankment type. These reservoirs are seldom covered. They present problems of algae growths, possible contamination, and water loss through evaporation, seepage, and leaks. Leaks and seepage may be reduced or stopped by lining the reservoir. More and more states are requiring operational reservoirs to be covered to reduce the hazard of possible contamination. Because earth-fill reservoirs are expensive to cover, fewer earth-fill reservoirs are being built.

Presently, reservoirs are built of standard or prestressed reinforced concrete or of steel. Although early concrete reservoirs were open at the top, most have been covered (see Fig. VI-1), and almost all new ones are built with a closed top.

Local topography may determine the type of reservoir to be built. In hilly areas reservoirs may be located at ground level as long as they are located higher than the service area. In areas that are relatively flat, elevated storage is usually built.

The cost of construction of any elevated storage is considerable. Sufficient additional usable storage (at reduced pressures) is available, however, through standpipe construction. The primary value of this additional storage is for fire protection. True, if power outage occurs, pumps will not operate, and water flows must occur by gravity, but most fire-fighting elements utilize gas-engine driven pumpers and therefore can utilize most all of available storage in standpipes. Standpipes require thick steel walls at lower reaches in order to withstand stresses, but the result is additional water storage capacity.

Frequently, to service small areas, utilities will pump water into a sealed tank filled partially with air. The air is compressed on top of the water and thereby

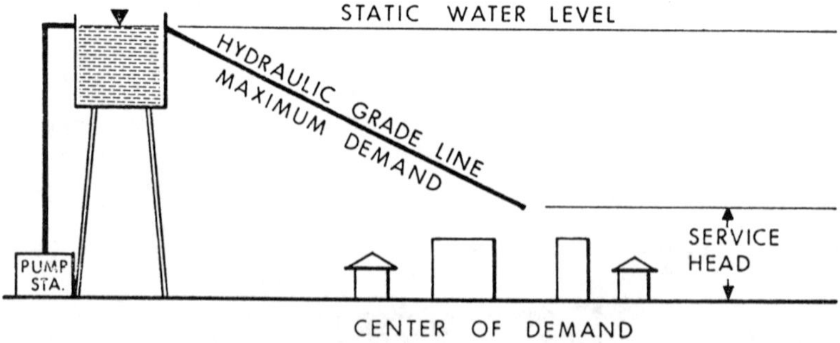

A. STORAGE AT SOURCE

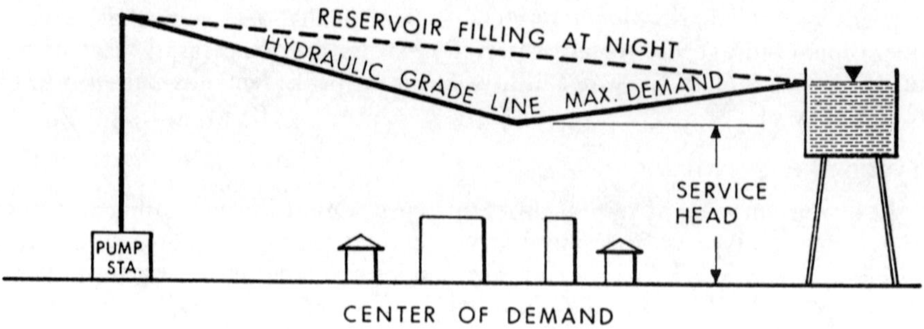

B. STORAGE OPPOSITE SOURCE

Fig. VI-1. Storage Deployment

provides the necessary pressure for servicing the area. This allows for less frequent use of the pump because it can be set to turn on only when the pressure drops below a certain point. This is called a hydropneumatic system.

An occasional utility will have wooden reservoirs. Although they are seldom built anymore, an operator should be aware of their existence. Wooden tanks must be carefully maintained because leaks in them can cause deterioration of areas surrounding the leak.

Many reservoirs are presently being built underground for appearance's sake or to allow the roof surface to be used for parking or recreation. Underground reservoirs should be carefully protected from seepage of surface- or ground-water fertilizer.

Purpose of Storage

Reservoirs "float" on the water system. This means that the flow into a reservoir, whether by pumps or some other means, is set at about the average demand for water. During times of the day when demand is low, the reservoir will be filling. During periods of peak or above-average demand, the reservoir will be emptying. Reservoirs, primarily, therefore average or equalize the demand on supply facilities.

Normally, distribution reservoirs are relatively small. Their sizes usually depend upon, primarily, the requirements for emergencies or contingencies such as fire flows and also the maximum demand for water and the maximum pumping capacity. The reservoir will usually be of sufficient size to contain at least a 4-6-hr supply of peak-period water. The height of the tank will largely depend upon the pressures needed in the distribution system.

Reservoirs allow pumps to be used at a uniform rate. This results in savings for the utility, which can operate with fewer pumps, and smaller water utilities also can contract for power during reduced demand periods when power rates are low. With treated-water-reservoir reserve it is also possible to have one- or two-shift operations rather than 24-hr-duty personnel. Without adequate reservoir capacity to handle periods of peak demand, pumping is required to maintain capacity and positive pressure throughout the system.

Water may be held in distribution reservoirs in order to provide an extended detention time for effective chlorination. The reservoir may also serve as a sand trap to allow the settling out of suspended solids in the water.

Reservoirs may be very useful if a utility is obtaining its water from wells. Waters from different wells can be mixed in a reservoir to provide water that will be of a consistent quality. Also, pumps may be automatically controlled to maintain the level in the reservoir, allowing for the use of low-capacity wells only during periods of sustained peak flow. This should result in a longer useful life for the low capacity wells.

Also a reserve capacity for fire emergencies is required by the Insurance Service Office (ISO).

Protecting against fire, averaging of peak and low flows, and providing pressure are the primary benefits of distribution reservoirs. These primary benefits lead to some secondary benefits that are also important to the utility.

Distribution reservoirs are often located in the outlying or low-pressure problem areas to provide the necessary pressure for the area during peak demand periods. They provide the necessary pressure by being either elevated, ground-storage located at the highest area, or ground storage with a booster station (whereby pumps provide the necessary pressure). (See Fig. VI-2.)

Maintenance of Reservoirs

Reservoir inspection and maintenance records should be kept for all reservoirs. Reservoirs should be inspected on a frequent basis. Any cracks or leaks should be repaired immediately. The overflow and air vents should be checked to ensure that they are adequately screened to prevent the entry of birds or insects, which could result in possible contamination of the water. The air vents should not be blocked in any way other than by screening because they are crucial in the filling or emptying of the reservoirs. If the vents do not allow air to escape fast enough, pressures can build up inside the tank during filling and cause extensive damage to the reservoir. Also, if air cannot enter to replace the water being drawn out, a vacuum in the reservoir can occur. This can also damage or collapse the reservoir.

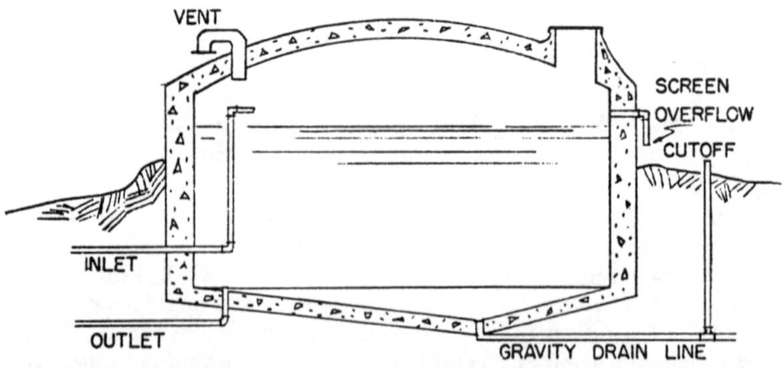

Fig. VI-2. Ground Storage Reservoir

The manhole for entry into the reservoir should be locked at all times, Also, the door on the manhole should curl over the lip of the manhole and not seat directly into it. This prevents water from washing in any foreign substances. The grounds surrounding the reservoir should be adequately fenced to prevent tampering with the reservoir and the grounds should slope away from the reservoir. This is especially important with elevated tanks since poor drainage around the foundations can lead to weakened support of the footing.

Tanks should be drained, cleaned, and disinfected as often as necessary to maintain the tank's integrity and the water's quality. Reservoirs normally have a floor or base that slopes to a drain. The drain should be valved and should not be connected to any sewer lines. Such a tie-in to a sewer line would be a cross connection that could leak to a hazardous contamination of the water supply through possible back flow of sewerage into the tank.

After the tank is drained it should be thoroughly washed. All silt at the bottom should be removed. Any growth of plant life between the high- and low-flow elevations should be removed. The growth of plant life is called littoral growth and could result in quality problems.

Occasionally the inside of a steel reservoir will require painting. When this occurs, the reservoir should be allowed to dry out completely and the reservoir should be cleaned. Care must be taken to remove all traces of rust or corrosion as any spots that are missed can seed the resumption of rusting and corrosion shortly after the reservoir is refilled. Reservoirs are usually painted with a bituminous paint. The paint should meet AWWA specifications to ensure that it will not be a source of poison or bad taste in the water supply. Many paints contain lead and are prohibited from use. The paint should be allowed to dry completely between coats and before refilling of the reservoir.

Reservoirs should be disinfected either before filling or during filling. Several methods exist.

The walls and other surfaces can be sprayed with a solution containing about 500 ppm (mg/1) of chlorine to provide immediate disinfection. This is a hazardous task that requires a man to enter the reservoir with an approved breathing apparatus and fully protective clothing. The person in the tank should be connected with another person outside the tank by a rope. This technique, although feasible, is extremely hazardous. One of the following techniques would be preferred.

Hypochlorites can be added by a portable hypochlorinator or in solution directly to the water. Liquid chlorine can also be applied through a chlorinator to the filling water. Whichever method is used, a 50 ppm (mg/1) residual should be maintained in the tank for a sufficient period to ensure the disinfection of the tank. This period of time will normally be at least 6 hr and preferably 24 hr. The water should then be drained and the reservoir put into service after samples indicate that it is ready for service.

Cathodic Protection

In any steel reservoir a natural flow of electric current takes place. Current in the form of positive metal ions flows away from an area of the steel reservoir (called the anode or anodic area) through the water (serving as an electrolyte and then back to another area of the reservoir (called the cathode or cathodic area).

When the current flows away from the anode area it also carries metal into solution. For each ampere of current flow, 20-24 lb of steel are removed each year. This is the cause of one of the most common forms of corrosion. In order to prevent this, cathodic protection systems are added to steel reservoirs. [Pipelines frequently have cathodic protection added also; the principle is basically the same (see Chapter VIII.)]

Because the corrosion occurs at the anodic areas only, the problem is solved by substituting anodes that are readily and economically replaceable. Anodes may be of different material, but the most common is an iron alloy. A DC power supply (usually a device that converts local AC to DC) provides a current flow to the anodes, which is discharged into the water. The current completes its circuit by flowing from the water to the reservoir structure or well, which is also connected to the power supply. This results in the entire reservoir structure being only cathodic (noncorrosive) — thus, the name "cathodic protection."

Cathodic protection systems can be valuable in preventing corrosion and should result in savings for cleaning and painting. They also reduce downtime. They must almost always be custom-designed for the job. Cathodic protection in a reservoir is only effective for those reservoir areas that are covered by water. Cathodic protection does not eliminate the need for high-quality coating protection, however.

Chapter VII
Pipe, Types, Uses and Protection

Introduction

This chapter is intended to provide the water-distribution operator with an understanding of the basics with regard to piping and should enable him to understand the choice of piping for his system. It should also explain the desirable qualities of pipe and the advantages and disadvantages of different varieties of pipe.

Features of Pipe

Pipe must have sufficient strength to resist a variety of forces. Pipe placed into service is usually covered by a layer of dirt or gravel. If laid in a street it must be able to withstand the load of passing vehicles. The pipe must also be able to resist damage from impact. The ability to withstand exterior forces such as these is a measure of the crushing strength of a pipe.

In addition to the external pressures, pipe is subjected to internal pressures from water: delivery pressure, surges, and water hammer. Water hammer is the tremendous force that is created by a sudden change in the water's velocity such as that caused by closing a valve too quickly or starting pumps too rapidly. Water hammer travels rapidly through the pipe and can cause extensive damage by rupturing pipe and tearing out valves.

The bursting strength refers to the ability of the pipe to withstand internal forces. When pipes are rated for either crushing or bursting strength, an extra margin is included as a safety factor.

Pressure ratings of pipe refer to the working pressure to be expected within a pipe. The pressure ratings take into account the internal pressure and a safety factor. Normally pipe is carefully chosen to ensure that its pressure rating is adequate to handle the pressures in a system. For this reason, personnel replacing a section of pipe should be careful that a new piece has a pressure rating equal to or greater than that of the replaced piece. Generally there are four classes of working pressure that are commonly encountered. They are 100, 150, 200, and 250 psi.

When a pipe is subjected to pressures either external or internal that exceed its ratings, it may rupture or be crushed. Other common types of breakage are shear and beam breakage. A shear break occurs when the earth shifts.

Beam breakage, which resembles a shear break, may occur when a pipe is unevenly supported along its length. Much as a teeter-totter breaks at its fulcrum when the ends are overloaded, a pipe that pases over a narrow and unyielding support point, like a rock or wall, may break when it is loaded like a beam. Other types of beam breakage may occur.

Other Performance Factors

The durability of the pipe refers to its ability to exist for a long time without deterioration. This is especially important to a water utility because pipelines are expected to provide service for many years. Frequent replacement of pipelines would mean higher rates for the consumer. Several factors are critical for a pipe to be considered durable. These factors can be assigned to the inner surface and the exterior surface of the pipe.

The inner surface of the pipe should not react with the water, should resist corrosion, and should have satisfactory flow characteristics. One of the factors generally affecting flow capacity is the friction of the water against the pipe. The amount of friction is dependent upon the roughness of the inner surface. The *C* factor of a pipe is a coefficient used in a pipe flow formulas and refers to the pipe's interior roughness. The smoother the interior surface the higher the *C* factor. New pipe used by water utilities usually has a *C* factor in the 100-150 range. A good durable pipe will maintain a smooth surface (high *C* factor) throughout its service life.

The exterior surface of the pipe must be corrosion resistant. If the piping is electrically conductive, then cathodic protection may be needed to avoid electrolytic decomposition.

Installation features are taken into consideration when a particular pipe is chosen. Some of the installation features that are important are weight, jointing, available sizes, and ease of tapping the line.

Weight is important because, in general, the heavier the pipe, the more expensive the installation.

Jointing is important not only for ease of installation but also with regard to any possible flexibility or deflection.

Size is critical because some types of pipe may not be available in the desired size.

The ease of tapping may be important, depending on whether or not other lines (such as service lines) are to be attached to the pipe.

Standards that are generally recognized as desirable requirements for piping have been established by AWWA, ANSI (American National Standards Institute) and ASTM (American Society for Testing and Materials). Most agencies use the guidelines to set their own specifications. These guidelines have been established by experimentation, testing, and experience in actual practice.

The remainder of this chapter is devoted to actual types of piping. The reader will find only the most commonly used piping materials described. However, many other types of material have been used.

Cast Iron Pipe — Grey & Ductile

There are two types of cast-iron pipe — grey cast iron and ductile iron.

Grey cast-iron pipe has been in use for several hundred years. Ductile-iron pipe is relatively new (since 1948) and is the predominant cast-iron pipe installed for water lines.

Grey cast-iron pipe is cast in several different ways. Presently a centrifugal casting process is used with sand-lined or metal molds. Ductile-iron pipe also is cast by a centrifugal process but with procedures additional to those for grey cast-iron pipe.

Grey cast-iron is tough, impervious, and easily tapped. Ductile-iron is similar but, because of its chemical composition, it is stronger, less rigid, and has structural properties more nearly resembling steel.

Cast-iron pipe is internally linedto prevent tuberculation. Tuberculation is the pitting and growth of nodules which increase friction and decrease the pipe's *C* factor. This type of pipe is usually lined with cement-mortar or with other available linings. Unlined pipe may develop twice the friction loss of lined pipe. External coatings to reduce corrosion are normally of a bituminous material.

Cast-iron pipe is produced in sizes ranging from 2 to 54 in. It may be used for transmission lines, distribution lines, or in-plant piping. This type of pipe, if properly coated and lined, should provide an extremely long service life. In certain soils special protection such as cathodic protection or wrapping the pipe loosely with thin polyethylene sheets may be needed.

Cast-iron pipe is normally joined with one of five joints: bell and spigot, flanged, ball and socket, mechanical and push-on.

1. The bell and spigot is the oldest joint. It is not used in new installations. It requires caulking material such as lead or lead substitutes, which should be sterilized before use to prevent bacterial growth.
2. Flanged joints are used only for aboveground installations.
3. Ball-and-socket joints allow for considerable deflection and are used in special situations — usually water crossings.
4. The mechanical joint is still in use. Bolts through a retaining ring compress the gasket between the bell and spigot end of the pipe. Joint deflection of a few degrees is possible.
5. The push-on joint is the most widely used and the newest form of jointing. It requires less skilled labor to install, provides a non-leak joint, permits joint deflection of several degrees and has an overall lower cost.

Asbestos-Cement (A-C) Pipe

Asbestos-cement pipe is made of asbestos fibers, silica sand, and cement. The asbestos fibers provide much of the strength of the pipe. Two types of A-C pipe are available — water cured and autoclaved; i.e., the method used to develop the strength of the cement in the pipe is via heat supplied respectively by water or steam. A-C pipe is also available in class-strength and transmission-strength designations, depending on how the pipe is to be used. Generally for water distribution, the pipe is produced in working pressure classes of 100, 150, and 200 psi.

A-C pipe is lightweight and is easy to install and tap. This type of pipe is very resistant to corrosion but can suffer deterioration caused by water and soil conditions.

Because of its polished interior, the pipe has a high *C* factor. However, for use with soft waters, it should be specified with a low limit of uncombined calcium hydroxide to prevent leaching.

A-C pipe is nonconductive and therefore can neither be electrically thawed nor located by an electrical pipe locator. However, nonconductivity eliminates electrolysis as a problem. Care must be taken to ensure that A-C pipe is properly bedded to minimize any flexural stresses.

A-C pipe is joined by sleeved couplings also made of asbestos-cement (see Fig VII-1). Couplings should be marked according to the pressure of pipe with which they are to be used. The couplings consist of a sleeve that has two interior rubber rings that provide for a watertight seal. The joint provides for a deflection of up to 6 deg with no leakage. The joint is normally jacked or pulled onto the pipe. The coupling requires no caulking and is relatively inexpensive. Except for the couplings, fittings for A-C pipe are generally made of cast iron.

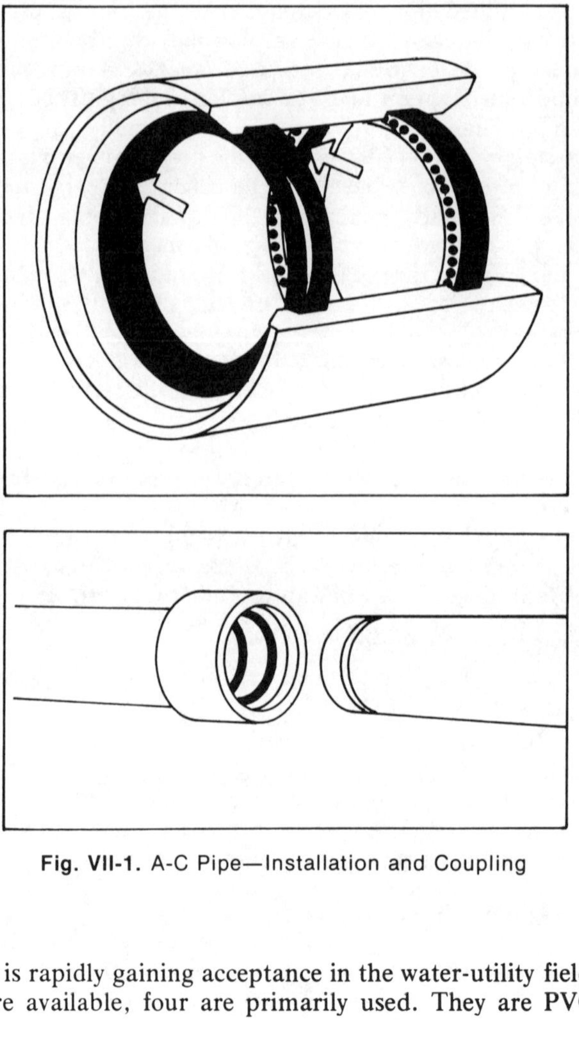

Fig. VII-1. A-C Pipe—Installation and Coupling

Plastic Pipe

Plastic pipe is rapidly gaining acceptance in the water-utility field. Although many types are available, four are primarily used. They are PVC (polyvinyl chloride), PE (polyethylene), PB (polybutylene) and ABS (acrylonitrile, butadiene, and styrene). ABS pipe use is diminishing because it lacks the strength of PVC pipe.

Plastic material may be thermoplastic or thermosetting. Thermoplastics can be softened and shaped more than once by the application of heat. PVC, PE, PV, and ABS are among the thermoplastics. Thermosetting materials cannot be softened after the initial formation of the material. Some types of fiber glass reinforced plastics are among the thermosets.

All plastic pipes possess a smooth interior surface and, thus, a high *C* value. Plastic pipe is lightweight and handles with ease. It is flexible in varying degrees, permitting deflection as the pipe is laid. Plastic pipe is resistant to corrosion and should provide a long service life.

Plastic pipe is produced in a variety of sizes. PE pipe is very flexible, is available in 100-ft rolls, and is usually available only in small sizes. PVC pipe may range from ½ to 36 in.* Some plastic pipe is available in long lengths, which thereby reduce the number of joints required. It is also easy to tap.

Plastic pipe is used for distribution lines and service lines. It has become extremely popular with those who irrigate because of its low cost and ease of installation and repair. Reinforced pipe is a lightweight semirigid pipe and is available in standard distribution sizes.

Fiber glass (layers of fiber glass bound by plastic resin) pipe is lightweight, flexible, and has a high burst strength. Fiber glass pipe is usually made in large sizes. When fiber glass pipe is exposed to sunlight, it needs a protective coating of sand-type material to prevent deterioration.

PE and PB varieties of pipe are usually joined by one of two different methods — insert fittings or flaring. Insert fittings require inside fittings at the pipe joints to make a connection. Flaring is similar in practice to flaring performed on copper tubing. However, with PE pipe, heat usually is necessary to make the plastic easily shapable. The compression-joint system is sometimes used with PE pipe. With this method an internal stiffener at the joint is required. The pipe is then placed into a fitting with a rubber gasket seal.

PVC pipe may be joined by a bell and spigot or by a solvent weld. In the bell-and-spigot method an O ring provides the necessary seal. In the solvent method a solvent that causes both pieces of pipe to be cemented together, creating a watertight seal, is used.

Plastic pipe should be bedded in the trench as are other pipe materials. Care should be exercised in backfilling to prevent any damage to the pipe. Plastic pipe exposed to sunlight requires a special coating to prevent deterioration. It should not be stored without a covering.

Steel Pipe

Steel pipe is strong and lightweight. It is easy to assemble, handle, and transport. However, it is more prone to corrosion and requires a lining or coating to maintain its service life. Three types of steel pipe are available: (1) fabricated, electrically welded for large pipe (4 - 144-in. diameter), made from rolled sheets of steel, electrically welded either longitudinally or in a spiral form

*AWWA Standard C 900-75 covers PVC pipe in sizes 4-12 in. diameter.

for strength; (2) mill type, either furnace or electrically welded, the most commonly used steel pipe (½ - 80 in.); and (3) seamless, available only in small sizes (2 - 4 in.).

Steel pipe coatings and linings are normally of a bituminous material or of a cement-mortar. Care must be exercised when handling and installing to ensure that the lining and coating are not damaged. Steel pipe may also have an epoxy lining. Materials used for coatings may include epoxy, galvanizing, mastic, or a protective wrap.

Large pipes are normally fitted with some type of vacuum relief. This is special protection to prevent a vacuum from occurring in the pipe, which would cause it to collapse inward. Steel pipe may also be provided with cathodic protection to prevent corrosion. (For diagrams of cathodic corrosion see Fig. VII-2.) Steel pipe may require special attention to prevent damage to protective coatings during shipping and handling.

Steel pipe may be joined in a variety of ways. Common joining methods are welding, compression couplings, push-on, mechanical couplings (Fig. VII-3), threaded couplings, and flanged joints. In these couplings, a resilient gasket forms the watertight seal.

Concrete Pipe

Concrete pipe is normally classified by the method used for reinforcement. Three types are: (1) steel cylinder, (2) prestressed or not prestressed, and (3) noncylinder not pretensioned. Concrete pipe is used primarily for transmission and distribution. It is available in sizes 12 in. and larger. In large sizes it is usually less expensive than any other kind of pipe. It is strong, durable, easily installed, and provides for a sustained carrying capacity. The disadvantages of

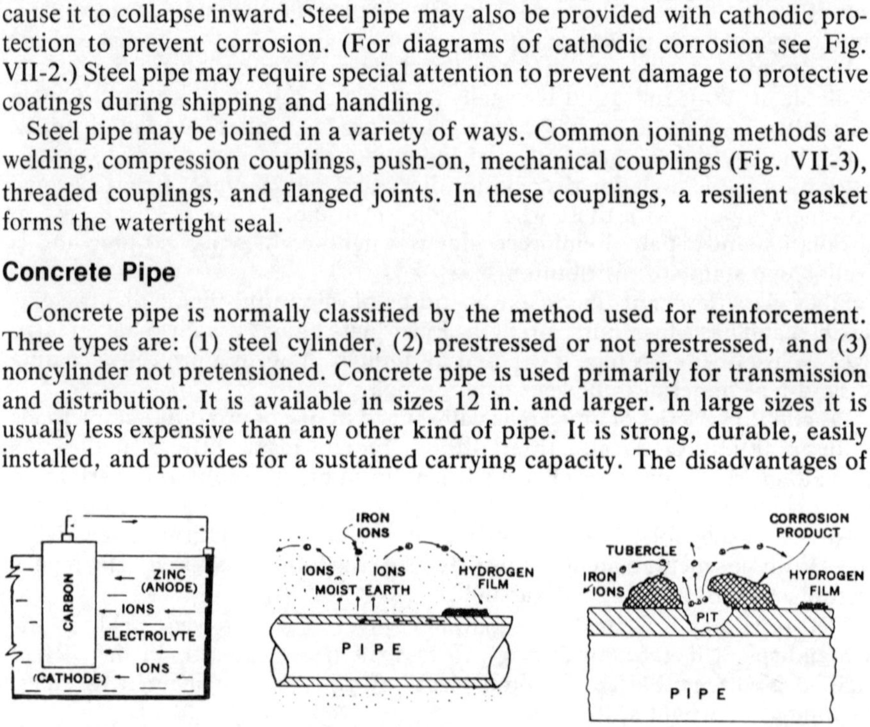

(a) BATTERY ACTION

(b) IONS IN MOTION START CORROSION

(c) CORROSION PROCEEDS WITH PITTING & TUBERCULATION

Principles of Galvanic Corrosion.

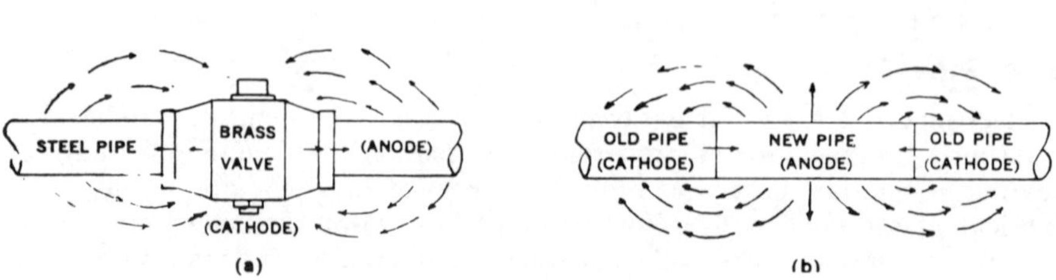

(a)

(b)

Corrosion Caused by Dissimilar Metals.

Fig. VII-2. Cathodic Corrosion

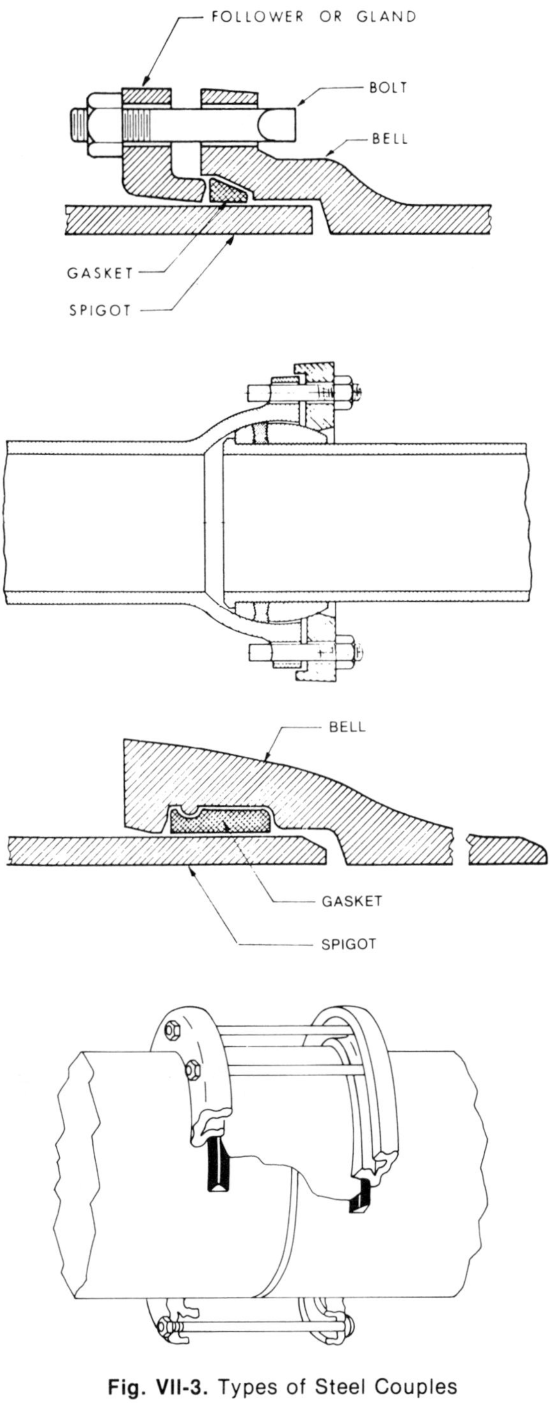

Fig. VII-3. Types of Steel Couples

concrete pipe are that the pipe is heavy and may be hard to tap because special equipment is needed.

Joining concrete pipe is normally accomplished by the use of a modified bell and spigot (see Fig. VII-4). The joints provide for some deflection. Bell-and-spigot type joints are normally filled with concrete 3etween the pipe lengths on the inside and outside.

Service Piping

Service piping provides the connection between the distribution mains and the individual service. The most important factor affecting the useful life of service lines is the ability of the material used to resist internal and external corrosion. To maintain the carrying capacity of service pipes and to prevent replacement because of deterioration, most service lines are copper, PVC, PE, PB, or, in special cases, reinforced plastic. Service lines may also be made of brass, cast-iron, black and galvanized wrought-iron and steel, and asbestos-cement. Some older service lines made of lead are still in existence, but lead lines are no longer acceptable and should be replaced.

Service lines are normally ¾ - 1½ in. Services are connected to the distribution mains through a shutoff valve called a corporation stop, or cock. In general practice most utilities own the service pipe from the distribution main to the meter and maintain it to this point. It is usually the responsibility of the homeowner to maintain the service line from the meter to the house (see "Service Installation" in Chapter X).

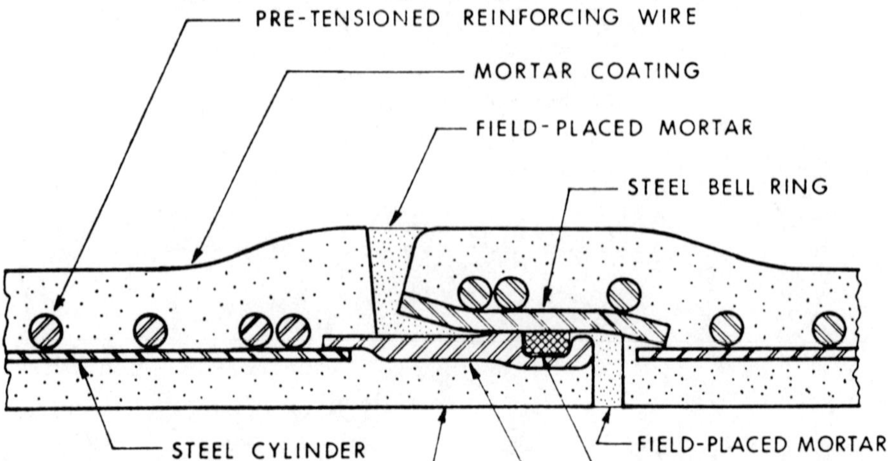

Fig. VII-4. Concrete Pipe Coupling

Chapter VIII
Pipe Installation and Maintenance

Introduction

Although some water operators are not directly involved with the subjects, a knowledge of pipe installation and maintenance is important in order for the operator to understand the workings of the utility. The operator has public contact with the water customer and may need to explain the utility's operation. Installations of new mains to service new customers and replacements of old pipe are generally understood and accepted by the public. However, installations of mains as general or localized reinforcements to increase the amount of water or pressure may not be understood.

Handling of Pipe

All pipe should be handled with care. Pipe normally receives a final inspection before leaving the factory. However, damage can occur in transit, and all pipe should be inspected when received. Check all materials, pipe, rings, gaskets and fittings received against a tally sheet.

Care must be exercised when unloading. Pipe should not be dropped or allowed to strike other pipe. Lined and coated pipe must be handled carefully to avoid damaging the protective material. If pipe is stacked it should be done in accordance with the manufacturer's directions. Belled pipe is normally stacked with the bells at opposite ends in each layer. Pipe should be stacked or stored off the ground to minimize contamination. Random lengths should be stored separately so that they can be located easily.

When the pipe is unloaded at the job site it should be placed as near as possible to where it is to be used to prevent excessive handling. Pipe is normally placed alongside the trench on the side opposite the excavated material. It should never be moved by the direct use of a metal blade or bucket such as on a bulldozer. All pipe, fittings, and gasket material should be protected from contamination, and kept as clean as possible.

Excavation and Installation

All buried water, sewer, gas, power, telephone, and cable TV lines and storm drains in the work area should be located and marked prior to excavating. Utility companies are usually willing to indicate the location of their facilities.

Trenches can be excavated in several ways, either by hand or machine. The latter includes the use of backhoes, trenching machines, and power shovels. A backhoe is often used in close quarters where the equipment operator can "feel" for underground obstructions such as other pipelines. The backhoe can also be used to place the pipe into the trench.

Trenches are dug to a specific level or grade. The depth of the trench will depend upon local conditions. Deep trenches are expensive but the pipe must be buried deep enough to provide protection from surface traffic, frost, water-channel erosion, and future changes in the ground-surface elevation.

Trenches should not be opened too far ahead of the work. In some localities it may be that no open trench is to be left exposed overnight. Trenches should be as narrow as possible, while providing room for pipe laying, backfilling, tamping, and other operations. Trenches should be shored and braced when conditions require it, or when specified by regulations of the Occupational Safety and Health Act. (See Fig. VIII-1.) The bottom of the trench should form a continuous bed for the pipe. (See Fig. VIII-2.)

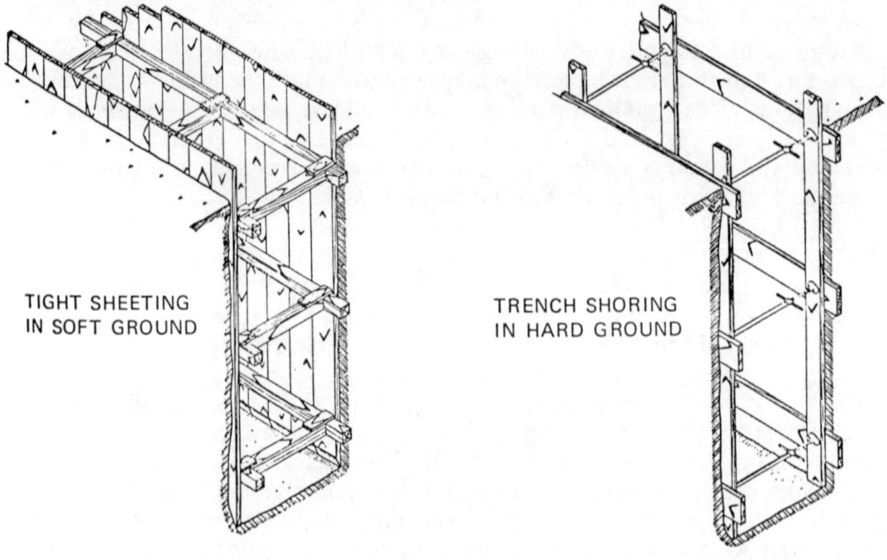

Fig. VIII-1. Shoring of Trenches

Pipe is sometimes laid on two earth pads at least 2 in. high and 12 in. long. Where rock is encountered trenches should be dug deeper than required, and then filled and tamped to grade with a bedding material such as sand or soil. Water should be kept out of the trench to avoid any possible contamination. Workmen should be provided with a latrine. At the end of each day or during long periods when pipe is not being laid, the end of the pipe should be plugged to prevent the entry of animals, foreign substances and water.

Sufficient space must be provided for joining pipe. Slip-on type joints require space for the bell. Occasionally, in some areas, because of dampness or other reasons, the soil will not provide adequate support of the pipe. When these con-

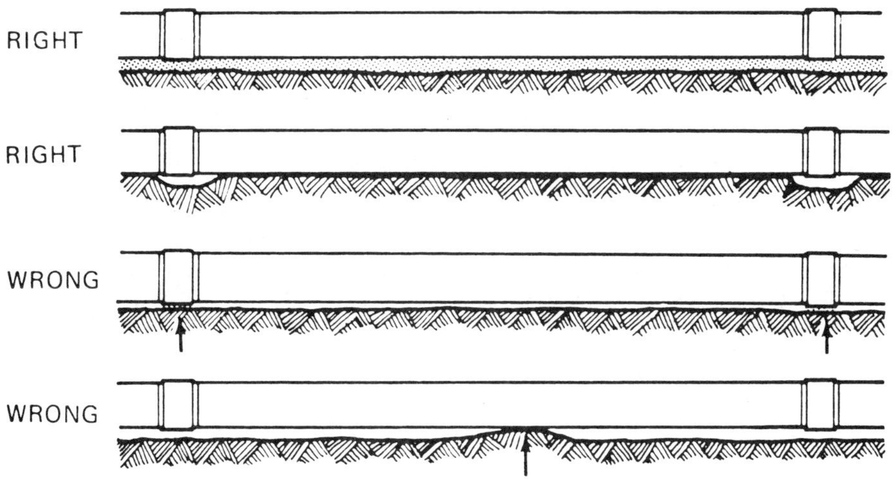

COUPLINGS—NEVER ALLOW COUPLINGS TO REST ON, OR TO SETTLE DOWN TO, ORIGINAL TRENCH BOTTOM.

PIPE—MAKE CERTAIN THAT PIPE BARREL IS GIVEN AN EVEN BEARING FOR ITS FULL LENGTH.

Fig. VIII-2. Bedding of Pipe—Right and Wrong

ditions are encountered, crushed rock, gravel, or pipe supports and stringers may be necessary.

Pipe should be inspected prior to being placed in the trench and all foreign material should be removed from it. The pipe may be swabbed with a hypochlorite solution prior to being placed. Pipe should never be allowed to drop into the trench. It should be lowered carefully into position.

Pipe for potable water should not be laid in the same trench with a sewer or nonpotable water line. Where a water line crosses a sewer line, the sewer line should be made of ductile cast iron with pressure-type joints for a distance covering 10 ft in each direction from the crossing, or be completely encased in concrete as required by local regulations. For added protection, water lines should be laid higher than sewer lines where in close proximity. Where a water line crosses beneath a railroad, the line should be laid inside a culvert pipe or steel casing that is at least 6 in. larger. The larger pipe will then provide the water line with protection from the extreme weight and vibration from the railroad.

Belled pipe is normally placed with the bells facing in the direction of the laying except on downhill slopes, where the bell faces uphill. For other types of joints the direction does not matter. Joints should not be deflected beyond manufacturer's specifications. All joints should be installed carefully. Joints often have a distinctive sound when correctly pulled together. Couplings that use rubber rings to provide a seal should be checked with a feeler gage.

All tees, bends, caps, plugs, hydrants, or other fittings that change the direction of flow or stop the flow should be restrained or blocked as shown in

Fig. VIII-3. This thrust restraint prevents the movement of the pipe that could be caused by surges, water hammer, or unbalanced pressure (see Chapter V). Movement of the pipe could result in leakage or disjointing.

Type "A" Blocking for 11¼°-22½°-30° Vertical Bends

		VB		S	d	L
Pipe Size Nom. Diameter-Inches	Test Pressure psi	Vertical Bend Degrees	No. of cu ft of Conc Blocking	Side of Cube feet	Diameter of Shackle Rods (2) inches	Depth of Rods in Concrete feet
4"	300	11¼	8	2	3/4"	1.5
		22½	11	2.2		2.0
		30	17	2.6		
6"	300	11¼	11	2.2	3/4"	2.0
		22½	25	2.9		
		30	41	3.5		
8"	300	11¼	16	2.5	3/4"	2.0
		22½	47	3.6		
		30	70	4.1	3/4"	2.5
12"	250	11¼	32	3.2	3/4"	2.0
		22½	88	4.5	7/8"	3.0
		30	132	5.1		
16"	225	11¼	70	4.1	7/8"	3.0
		22½	184	5.7	1 1/8"	4.0
		30	275	6.5	1 1/4"	
20"	200	11¼	91	4.5	7/8"	3.0
		22½	225	6.1	1 1/4"	4.0
		30	330	6.9	1 3/8"	4.5
24"	200	11¼	128	5.0	1"	3.5
		22½	320	6.8	1 3/8"	4.5
		30	480	7.9	1 5/8"	5.5

Type "B" Blocking for 45° Vertical Bends

		V.B.		S	d	L
Pipe Size Nom. Diameter-Inches	Test Pressure psi	Vertical Bend Degrees	No. of cu ft. of Conc Blocking	Side of Cube feet	Diameter of Shackle Rods (4) inches	Depth of Rods in Concrete feet
4"	300	45	30	3.1	3/4"	2.0
6"			68	4.1		
8"			123	5.0		
12"	250		232	6.1	3/4"	2.5
16"	225		478	7.8	1 1/8"	4.0
20"	200		560	8.2	1 1/4"	
24"			820	9.4	1 3/8"	4.5

Type A Blocking

for 11¼°-22½°-30° Vertical Bends

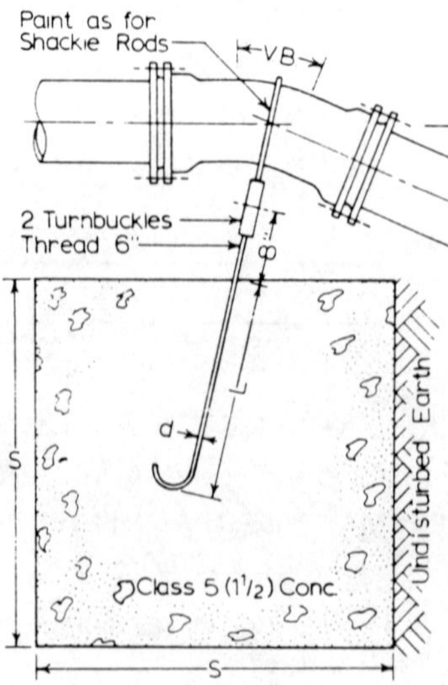

Type B Blocking

for 45° Vertical Bends

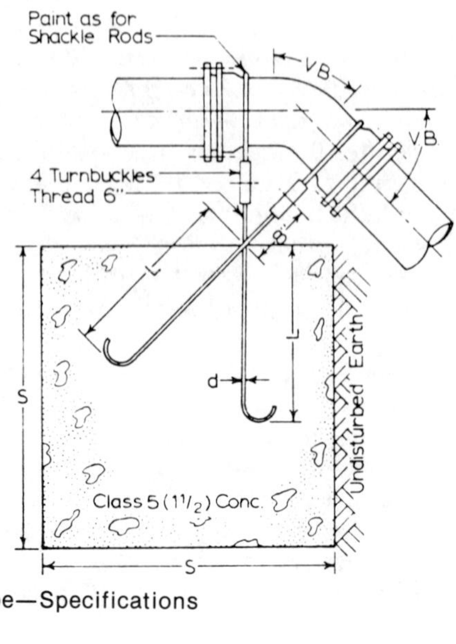

Fig. VIII-3. Blocking of Pipe—Specifications

There are different types of trench conditions which may be used by designers. Some call for no compaction of the backfill material, and others specify varying amounts of compaction around the pipe and above it. If compaction is required it should be accomplished in stages. (See Fig. VIII-4.) The complete tamping of the soil around the pipe should provide uniform support and bearing for it. Joints may be left exposed until the pipe has been hydrostatically tested (tested under water pressure). After such testing the remainder of the trench can be backfilled. NOTE: Empty lightweight pipe can float out of its bedding if rain water enters the ditch.

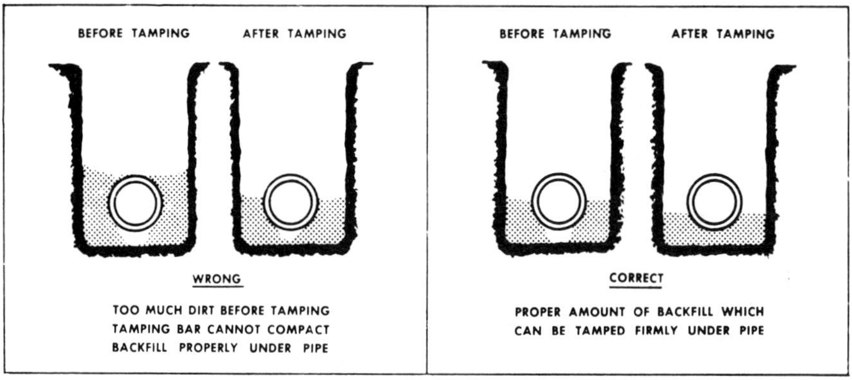

Fig. VIII-4. Tamping Procedures

The remaining backfill above the pipe can be tamped by hand or machine, or consolidated by applying water, depending on convenience and practicality. Trenches in roads or where sidewalks are to be located should be compacted to a density sufficient to prevent settling after the road or sidewalk has been placed.

Trenching is often avoided by jacking the pipe through the soil, but this is usually possible only for pipe 4-in. or smaller. The jacking method requires only two excavations—one where the pipe starts and one where it ends. Other methods, such as boring or tunneling, can be used to avoid trenching and are useful for large diameter pipe.

Safety

Safety is critical during pipeline installation and repair. New regulations under the Williams and Steiger Occupational Safety and Health Act (OSHA) define clearly what is permissible and what is not. The reader should refer to Chapter XIII in this text. Topics in that chapter, which are important during pipeline installation, include trench construction, pipe handling, barricades, warning signs, and traffic control.

Hydrostatic and Leakage Tests

After the trench has been partially backfilled the pipe should be tested for leakage. The usual procedure for leakage testing is to fill the pipe — either the complete pipe or sections of it — slowly with water, allowing the air to escape through corporation cocks or by some other means. When the pipe is full of water it should be allowed to sit idle for up to 24 hr, after which pressure in the line should be increased to 1.5 times normal operating pressure or 150 psi, whichever is larger, for 1-2 hr. Observations can then be made for any leakage, by sight or by the use of a meter to detect any flow into the pipe. Some leakage is to be expected. However, joints that leak excessively should be relaid. Some formulas are available to calculate the amount of allowable leakage. One such formula is

$$L = \frac{ND\,(P)^{1/2}}{3700}$$

Where L = allowable leakage in gallons per hour
N = number of joints in section tested
D = nominal pipe diameter in inches
P = average test pressure in pounds per square inch

Disinfection of the New Main

Before the new main can be put into service the line must be flushed and disinfected. Flushing is necessary to remove any dirt or foreign material from the pipeline. Flushing should be done with sufficient velocity to cleanse the pipe's interior thoroughly.

After flushing, the line should be disinfected. Lines may be disinfected with liquid chlorine or sodium or calcium hypochlorites. State health agencies should be contacted for acceptable methods. The three most generally recognized methods for accomplishing disinfection are the continuous feed, slug, and tablet methods (described in the appendix to this chapter). Disinfection should be accomplished by experienced workmen.

Cleaning of Pipe

Piping frequently builds up deposits that must be removed to maintain the carrying capacity of the line. Sediment deposits may be removed by semiannual or annual flushings. When flushing is not adequate, a spring-loaded instrument (sometimes called a pig) with steel scrapers can be forced through the line by water pressure or by a winch. Sometimes several spongy balls of increasing size are sent through a pipe for cleaning. During cleaning, temporary service should be provided to customers, with pipe laid above ground.

Tuberculation and scale deposits can be removed by the use of mechanical or hydraulic scrapers. The scrapers remove the tuberculation and existing coatings by a honing action. Care must be exercised after each scraping to ensure that deposits removed are flushed away and not allowed to enter service connections.

After scraping, pipes can be lined in place by the application of a cement mortar. A thin lining of ¼ in. or less is preferred in order not to reduce the inner diameter of the pipe too seriously. There are three ways of applying the lining: centrifugal method, reinforced centrifugal method, and mandrel method. Pipe that has been scraped is almost always lined afterward because, after scraping, tuberculation will occur at a greater rate than before. Cleaning and lining of pipe should result in lower pumping costs, increased volume, decreased pressure drop, and, therefore, increased water pressure to the customer. All connections should be checked and unplugged after lining.

Chemical Treatment of Water Mains

The use of some chemicals can protect the pipe against reduced flow capacities. Algae and slime growth can be controlled by chlorine. Raising the pH of an aggressive water decreases its aggressiveness and lays down a coating of calcium that protects the pipe. Sodium hexametaphosphate may be used to combat red water.

Emergency Repairs

Emergency repairs, such as for pipeline breaks, or leaks, provide serious challenges to any utility. All utilities should have a plan of action for emergencies. Spare parts and adequate tools and equipment should be available.

Before any repairs can be started the leak must be found. There are many ways to locate leaks, such as detecting wet spots or sunken spots in the street and unusually green weeds, using sounding rods and electronic amplification of the sound of running water, and observing neighborhood sewer flows. Any suspicious seepage or standing water can be analyzed to ascertain if it is leakage from a water main.

If possible, a leak should be repaired without shutting off the pressure, not only because of inconvenience to the customers but because contaminated water may drain back into the empty pipe. (Retaining pressure within the pipe prevents this.) Repair clamps or sleeves may be placed on a pipe alongside a leak, then slid over the leak, and tightened into place.

When a leaking pipe is excavated, all water should be pumped from the trench. After the pipe is repaired the isolated section should be flushed for a 24-hr period maintaining not less than 25 ppm (mg/1) chlorine (see appendix of this chapter, and AWWA Standard C601) before returning the pipe to service.

Repairs should also be hydrostatically tested. In the case of emergency repairs when the line may only be out of service for a short while, the slug of chlorine for disinfection may be increased to 500 ppm (mg/1). Time allowed for disinfection can then be reduced to 30 min. The highly chlorinated water must then be flushed from the line. (See the appendix of this chapter for disinfection of water mains.) A highly concentrated chlorine solution should not enter a customer service line.

Service Lines

Service lines and the installation of meters will be covered in Chapter X, "Meters and Services."

Cross Connections

The water operator should always be alert to avoid the chance of any cross connection (connection to nonpotable water). This is especially true when pipe is being installed. To eliminate any possibility of pollution, there should be no physical connection between an approved supply and one not approved.

Sometimes water operators are guilty of making cross connections. A water hose may be dropped into a drum to dilute chemicals, or perhaps a hose is placed in a plugged sewer drain. A sudden loss of pressure could cause these hoses to siphon the chemicals or sewage.

The operator should watch for reservoir drains that are sometimes connected to the sewer system. He should also look for abandoned pipelines that might be connected to the distribution system through a valve that might accidentally open during a period of pressure loss.

Sometimes liquid fertilizer is injected into irrigation water before it is pumped to a higher elevation for use. Pump failure would allow this contaminated water to flow back into the water main.

Industrial plants often maintain direct connections of fresh makeup water to tanks containing water-chemical mixtures. Many times these connections are made after initial construction and inspection.

Appendix
Water Main Chlorination

Continuous Feed

In this method, water from the distribution system or other approved source and the type of chlorine selected are fed at a constant rate into the new main at a concentration of at least 50 ppm (mg/l) available chlorine. A properly adjusted hypochlorite solution injected into the main with a hypochlorinator, or liquid chlorine injected into the main through a solution-feed chlorinator and booster pump, may be used. The chlorine residual should be checked at intervals to ensure that the proper level is maintained. Chlorine application should continue until the entire main is filled. The water should remain in the pipe for a minimum of 24 hrs, during which time all valves and hydrants along the main must be operated to ensure their proper disinfection. Following the 24-hr period, no less than 25 ppm (mg/l) chlorine residual should remain in the pipeline.

Slug Method

In this method a continuous flow of water is fed with a constant dose of chlorine (as in the previous method) but with rates proportioned to yield a chlorine concentration of at least 300 ppm (mg/1). The chlorine is applied continuously for a period of time to provide a column of chlorinated water that will contact all interior surfaces of the main for a period of at least 3 hr. As the slug passes tees, crosses, and bends, proper valves must be operated to ensure their disinfection. This method is used principally for large-diameter mains where continuous feed is impractical.

Tablet Method

Because this method precludes preliminary flushing, it is considered the least satisfactory, although it is commonly used for small-diameter mains. Calcium hypochlorite tables are placed in each section of pipe, hydrants, and other appurtenances. These tablets should be placed on the top of inner walls of the pipe with approved adhesives. For the most uniform disinfection, the tablets should be placed in the pipe during installation. The main should then slowly be filled with water at a velocity of less than 1 fps to prevent washing of the hypochlorite to the end of the main. The final solution should have a residual of at least 50 ppm (mg/1) and should remain in contact for a minimum of 24 hr.

Regardless of the method used, it is necessary to make certain that backflow of the strong chlorine solution into the supply doesn't occur. After the prescribed contact period the chlorinated water should be flushed to waste until the remaining water has a chlorine residual approximating that throughout the rest of the system. Bacteriological tests as prescribed by the authorities should be taken. If the results fail to meet minimum standards, the disinfecting procedure must be repeated and the results tested again before placing the main in service. One should remember that too much chlorine for an excessive time can damage brass and other fittings.

Chapter IX
Pipeline Appurtenances

Introduction

This chapter is intended to provide a basic understanding of pipeline appurtenances. Appurtenances include fittings, valve boxes, valves and fire hydrants. For additional information refer to the sources in the bibliography.

Valves

To the water operator in the field it often seems that there are many different types of valves because valves are classified in many different ways: by use, by design, and by the material from which the valve is made. In this section valves will be discussed by design and then by applications. Valves can be made of any type of material that is needed for a special application.

Valves are produced in many different designs. The four most common types encountered by a water operator are diaphragm, globe, rotary, and slide (Fig. IX-1).

A diaphragm valve has a flexible piece inside the body of the valve, which can be adjusted up or down by an attached stem. The piece can be made from rubber, leather, or some flexible material that is attached to the perimeter of the valve. When depressed the piece serves to stop or regulate the flow of water.

Globe valves have a flat or tapered disc of rubber or leather that is raised or lowered (as in the common home faucet) onto a seat. When the disc contacts the seat the flow of water through the valve is stopped. Other globe valves are of a needle design. They have a cone-shaped piece protruding into the waterway to restrict or stop the flow. A needle valve is satisfactory for precise adjustments of small flows. Globe valves can be designed for straight, angle or **Y**-pattern flows. They are sometimes controlled by a diaphragm that controls the valve's position with pressures using pressure differentials across the diaphragm, making them suitable for automatic control.

Rotary valves include ball, butterfly, cone, and plug types. Ball valves have a round sphere inside them. The sphere has a passageway which can be rotated so that water can flow through the valve, be restricted, or stopped. The sphere is positioned between resilient seat washers that prevent leakage. Cone valves are designed so that a cone shaped element can be rotated within the valve body. The cone piece has an opening through it, which provides for full flow when the valve is completely open. The cone may also be turned to restrict or shut off the flow. Plug valves, similar to cone and ball valves, have a cylinder with a hole,

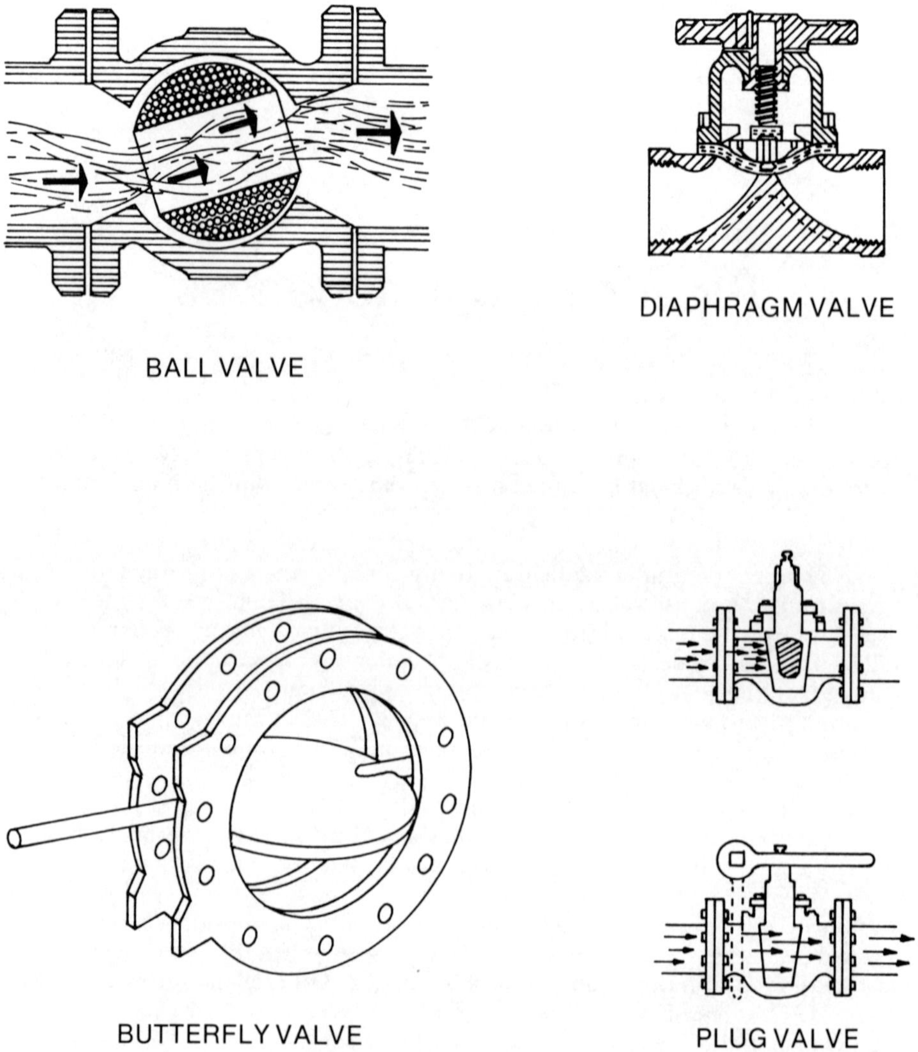

Fig. IX-1. Valve Designs

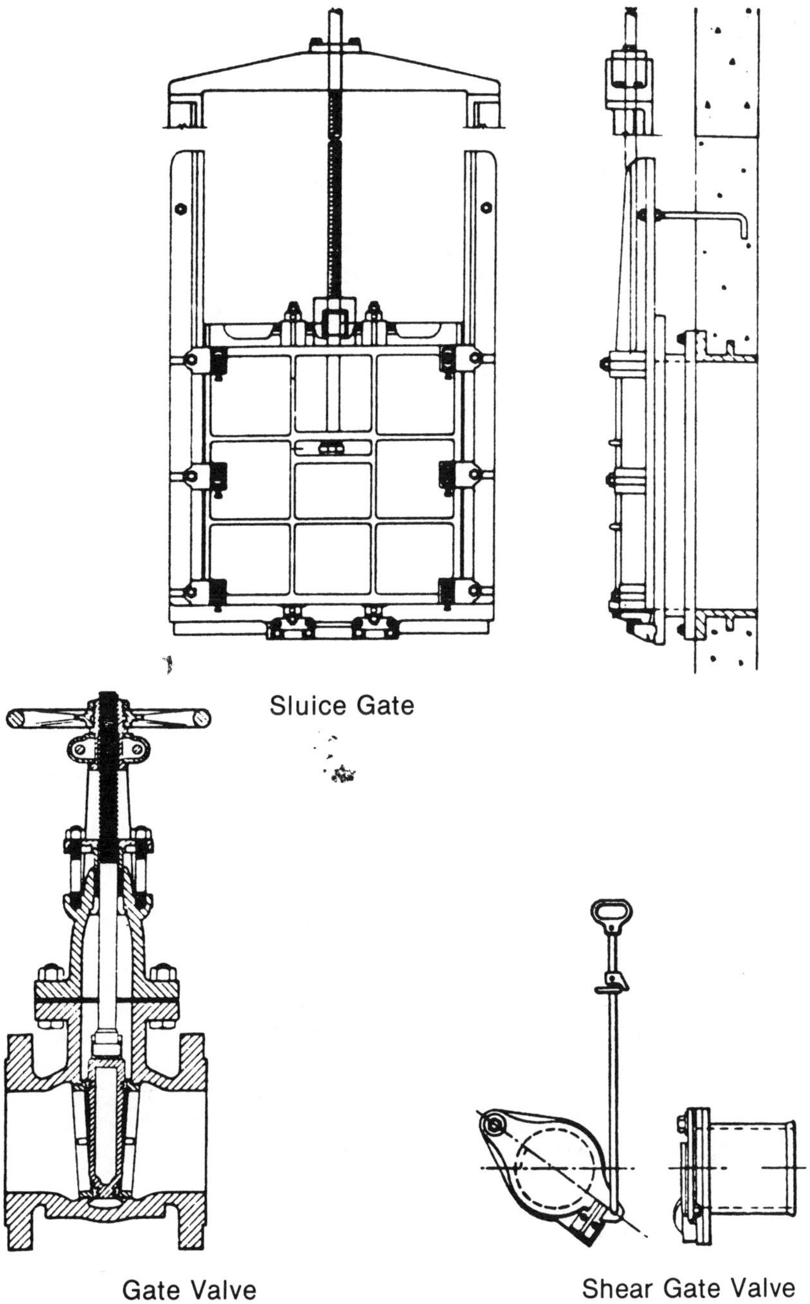

Fig. IX-1. Valve Designs

which can be turned to open, restrict, or close the flow. Butterfly valves are designed so that an internal disc can be turned either to block or allow the flow of water. The disc is connected to a shaft that extends into or through the valve body. Around the edge of the disc is a machined edge that seals against a gasket in the valve body. Rotary valves usually have a 90 deg rotation movement with a notch, arrow, or indicator to show valve positions.

Gate valves, sluice gates, and shear gates are types of slide valves. A gate valve has a disc that rises from or drops into the flow and moves in a channel that keeps it aligned. Discs in gate valves can be single disc, double disc, wedge, nonwedge or knife type. The gate-valve stem may or may not rise relative to the body of the valve as the disc is raised. If the valve stem rises above the body, the valve opening can be judged by the exposed stem. Because the disc can be raised completely out of the water, gate valves cause little reduction in flow when installed in a pipeline (see Fig. IX-2).

Sluice gates are usually rectangular and are raised or lowered to control flows. The gate moves in a frame built into a dam, weir, bulkhead, or channel. Generally the design is such that the pressure of the water behind the gate serves to seal it.

Shear gates are valves that have a plate covering the valve opening. The plate is rotated on a pin at one side of the seat and slides over the opening. It is moved by a rod that is attached opposite the pin. When the plate is closed a wedge forces it against the end or seat of the valve.

Valve Applications

Valves have many uses in a water distribution system. Diaphragm, slide, rotary, and globe valves can be used to provide starting and stopping functions. Globe, diaphragm, and rotary valves are usually used to regulate flows; some slide valves can also be used for that purpose.

Pressure regulating valves reduce water pressure by restricting the flow. Pressure on the downstream side of the valve regulates the amount of flow permitted through it. This type of valve is usually of the globe design with a spring-loaded diaphragm that sets the amount of the opening. As downstream pressure is exerted against the diaphragm the spring is compressed, moving the valve element toward the seal, thereby limiting the flow. If the downstream pressure drops the spring will open the valve element and provide more flow.

Air in water lines collects at peaks and at high spots. These pockets of air can increase the resistance to flow by 10-15 per cent. An air-lock condition might stop the flow of water completely. Vacuum conditions can occur in pipelines during draining of the line, which can cause the collapse of some pipelines. To alleviate these problems *air and vacuum relief* valves should be installed. Air and vacuum relief valves use a float to operate. When the water level is low in the pipe or when enough air accumulates in the valve, the float drops and allows the entry or exit of the air. When the water level rises, the float closes the opening and provides a watertight seal. On some models a small quantity of water is permitted to escape; therefore adequate drainage should be provided. If the air and vacuum relief valves are installed in a vault there should be adequate venting.

Properly designed air and vacuum-relief valves function in a way that pre-

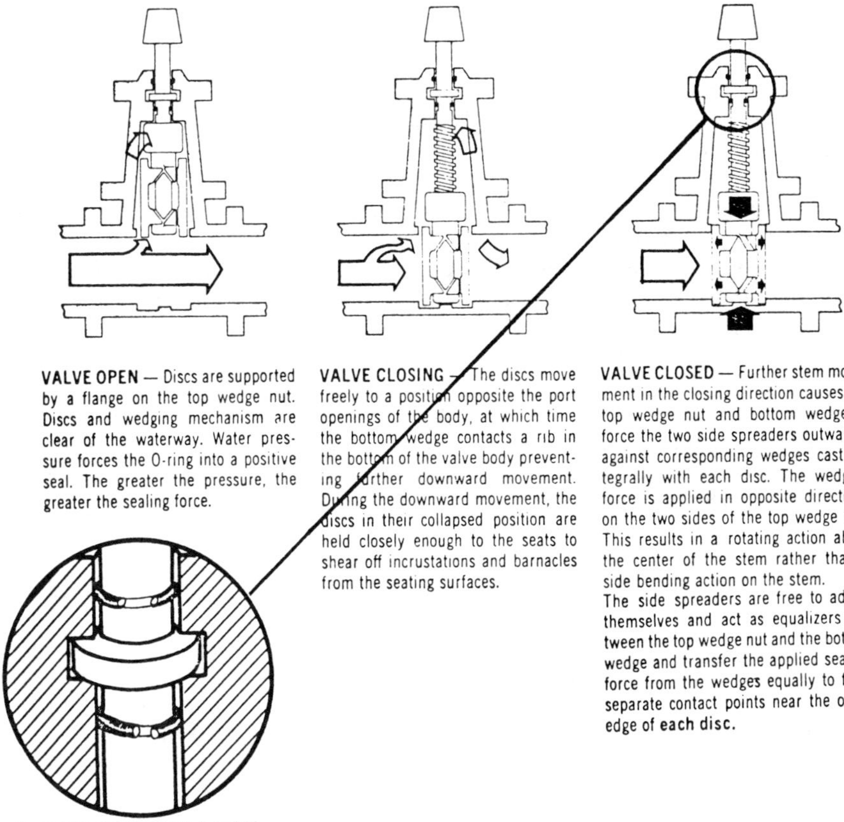

VALVE OPEN — Discs are supported by a flange on the top wedge nut. Discs and wedging mechanism are clear of the waterway. Water pressure forces the O-ring into a positive seal. The greater the pressure, the greater the sealing force.

VALVE CLOSING — The discs move freely to a position opposite the port openings of the body, at which time the bottom wedge contacts a rib in the bottom of the valve body preventing further downward movement. During the downward movement, the discs in their collapsed position are held closely enough to the seats to shear off incrustations and barnacles from the seating surfaces.

VALVE CLOSED — Further stem movement in the closing direction causes the top wedge nut and bottom wedge to force the two side spreaders outwardly against corresponding wedges cast integrally with each disc. The wedging force is applied in opposite directions on the two sides of the top wedge nut. This results in a rotating action about the center of the stem rather than a side bending action on the stem.
The side spreaders are free to adjust themselves and act as equalizers between the top wedge nut and the bottom wedge and transfer the applied seating force from the wedges equally to four separate contact points near the outer edge of each disc.

Fig. IX-2. Gate Valve Principle

vents large volumes of air from blowing the float into a constant closed position. Air and vacuum-relief valves can be combined as one unit, which is normally preferred for its economy.

Blowoff and drain valves are usually placed at low spots in water lines. These on-off type valves are used to dewater lines for repairs or to blow off any sediment that accumulates at low points.

Sectionalizing valves are gate valves that are installed at various points in the water distribution grid. Normally two valves are placed at an intersection of pipelines to permit the isolation of areas under repair or areas of contamination. Adequate placement of sectionalizing valves prevents large areas of service from being without water during main repairs.

Back-flow prevention or cross-connection control can be obtained through the use of specific valve units. These units provide for normal flow, but when pressure levels drop or back-flow occurs these valves close off the flow of water (Fig. IX-3). Globe or swing designs (Fig. IX-4) are normally used for this purpose. A double check-valve arrangement is generally used in back-flow prevention. A check valve is a type of valve designed to permit flow in only one direction. When this type of arrangement is used, a means must exist for testing the effectiveness of the check valves. This is accomplished by installing petcocks between valves to check for reverse leakage (Fig. IX-5).

A water utility may desire to maintain the water level in a reservoir at a certain elevation, which can be accomplished through the use of an altitude valve. This type of valve is generally operated by the water pressure or a float and is usually a globe valve.

There are many other valve applications in a water utility. The operator should attempt to learn the function of the valves in a system and learn to recognize their design characteristics.

Valve Operation

Valves can be operated either manually or by power operation. Manual operation of valves can be accomplished by the direct turning of the valve stem or by a geared arrangement to permit easier turning of the stem of large valves. The power operation of valves can include a motor or a hydraulic or air piston to operate the stem.

When one operates valves two types of problems can occur: cavitation and water hammer. Both of these problems can be costly to a water utility but can be avoided through proper operation and design.

Frequently in the operation of valves, conditions cause a partial vacuum or void to occur on the downstream side of a valve. These voids will fill with low-pressure vapors from the water. When these pockets implode or collapse they create a mechanical shock that causes pockets of metal to break away from the valve surface. A noisy or vibrating valve may be an indication that cavitation is occurring, which will eventually result in leaks and a valve that is unsuitable for service.

Water hammer is caused by closing valves too quickly. When water is suddenly stopped, shock waves are generated, which cause large pressure increases throughout the system. These shock waves travel quickly and can cause extensive damage. Water hammer during operation of valves can be prevented if the operator remembers to close valves and hydrants slowly.

Water hammer and cavitation are also problems in the operation of pumps. These subjects are discussed again in Chapter XI.

Joints & Fittings

Valves are inserted into a pipe system using a variety of couplings. A few varieties are shown in combination with various valve designs in Fig. IX-6.

Fittings as Joinings. These have a variety of uses for a water utility. They can connect similar pipe, connect two different sizes of pipe, change the direction of flow, and stop the flow.

Fittings are available in most materials and sizes. Some are designed for the connection of two dissimilar piping materials. Fittings are also available for connecting pipes with different types of ends.

Fittings that are used to connect pipe include clamps, sleeves, couplings, adapters, and expansion joints. Fittings that reduce the size of pipe are called reducers and include reducing flanges, tees, Y-laterals, and crosses. Elbows, tees,

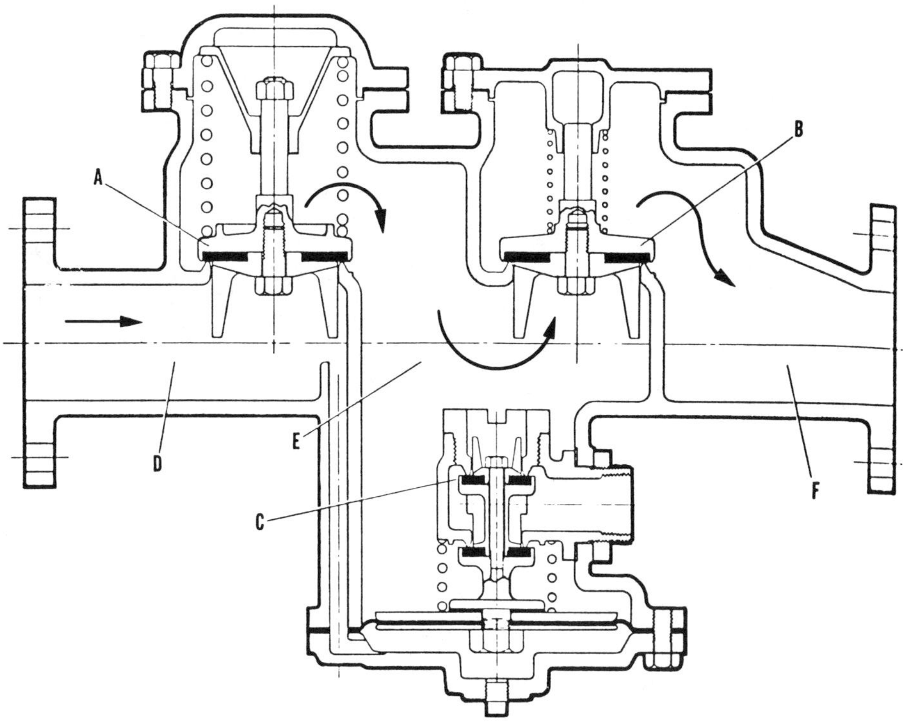

This device operates on the principle that water will not flow from a zone of lower pressure to one of higher pressure. It provides maximum protection against backflow caused by both backpressure and backsiphonage. The device consists of two spring-loaded check valves (A and B) and a spring-loaded, diaphragm-actuated, differential-pressure relief valve (C) located in the zone between the check valves. The first check valve (A) causes all water passing through it to be automatically reduced in pressure by approximately 8 psi. The second check valve (B) is lightly spring loaded and forms the "double check" feature of the device. It acts to prevent unnecessary drainage of the domestic system in case a backflow condition occurs.

Fig. IX-3. Reduced Pressure Backflow-Preventer

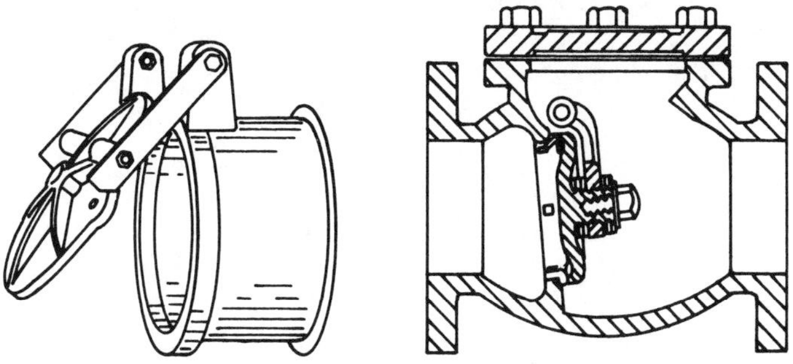

Fig. IX-4. Globe (Swing) Valves

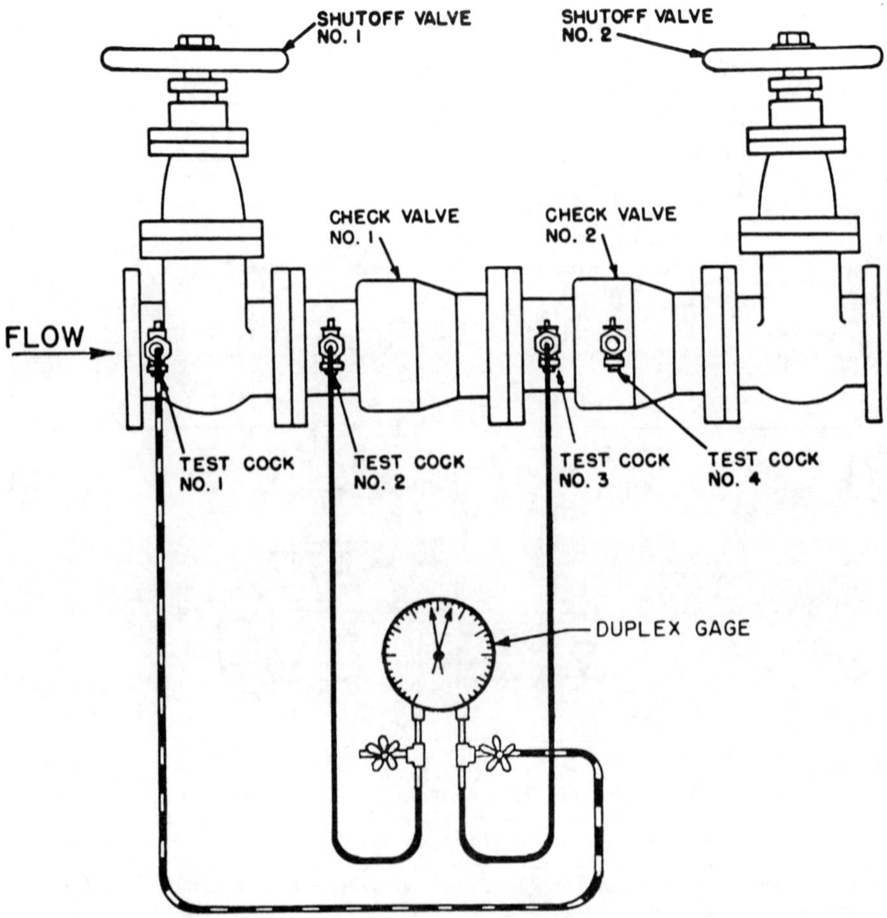

Fig. IX-5. Setup for Checking Out Backflow Preventer

and bends are used to change direction of flow. For stopping flow there are caps, plugs, and blind flanges (a flange with no opening). Fittings for tapping pipe include tapping sleeves, tapping crosses and tapping clamps.

Standards for fittings are available from the American Water Works Association.

Valve Boxes and Vaults

Valve boxes are similar to curb boxes (see Chapter X). They are used to provide access to buried valves. Through the use of a valve key (a portable extension of the valve stem), the operator is able to turn the valve stem from the top of the structure. Normally, valve boxes are adjustable so that the top of the box can be set at ground level.

Vaults are normally used for larger valves where access for maintenance operations is necessary. They may vary in size and usually provide enough room

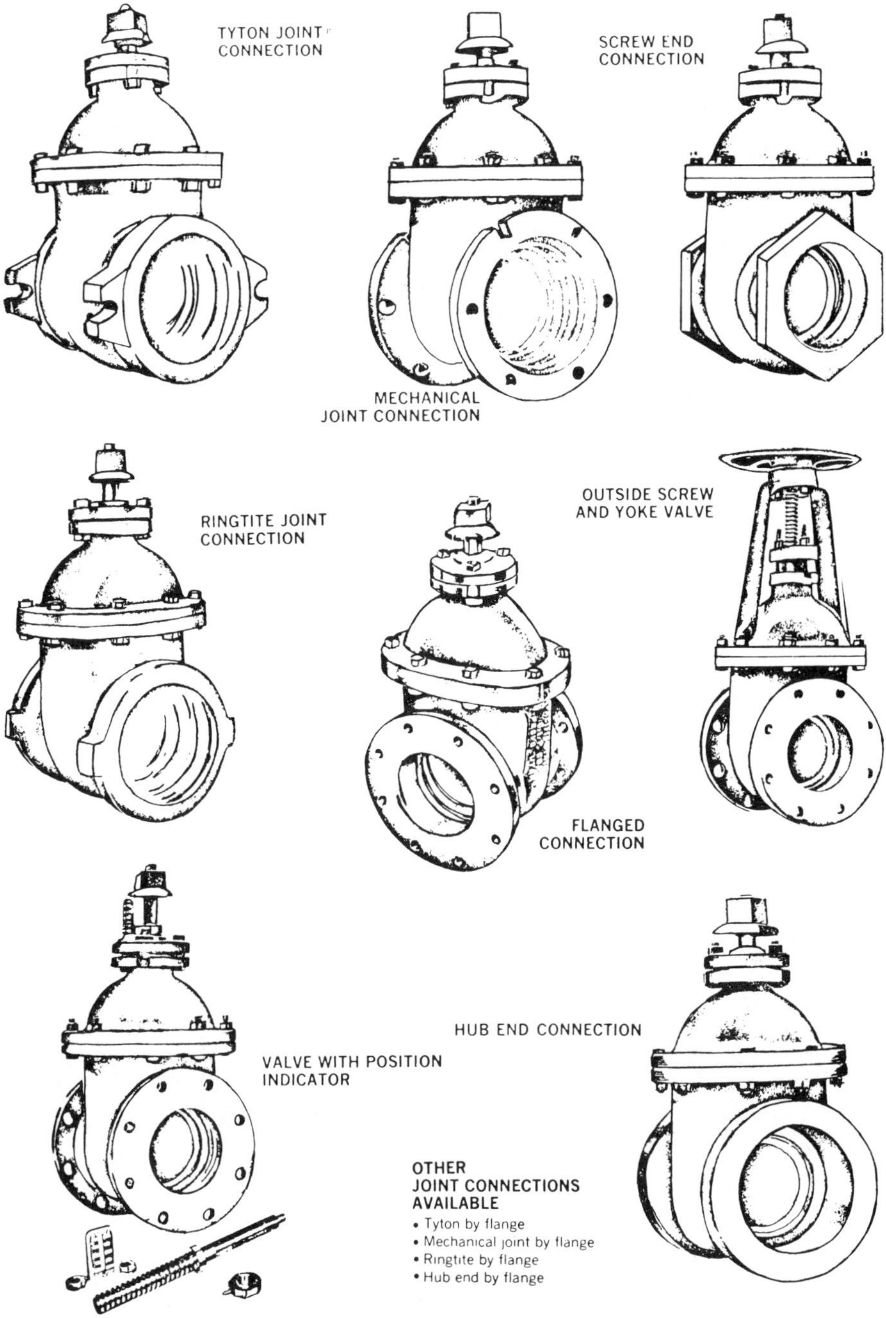

Fig. IX-6. Valve Couplings

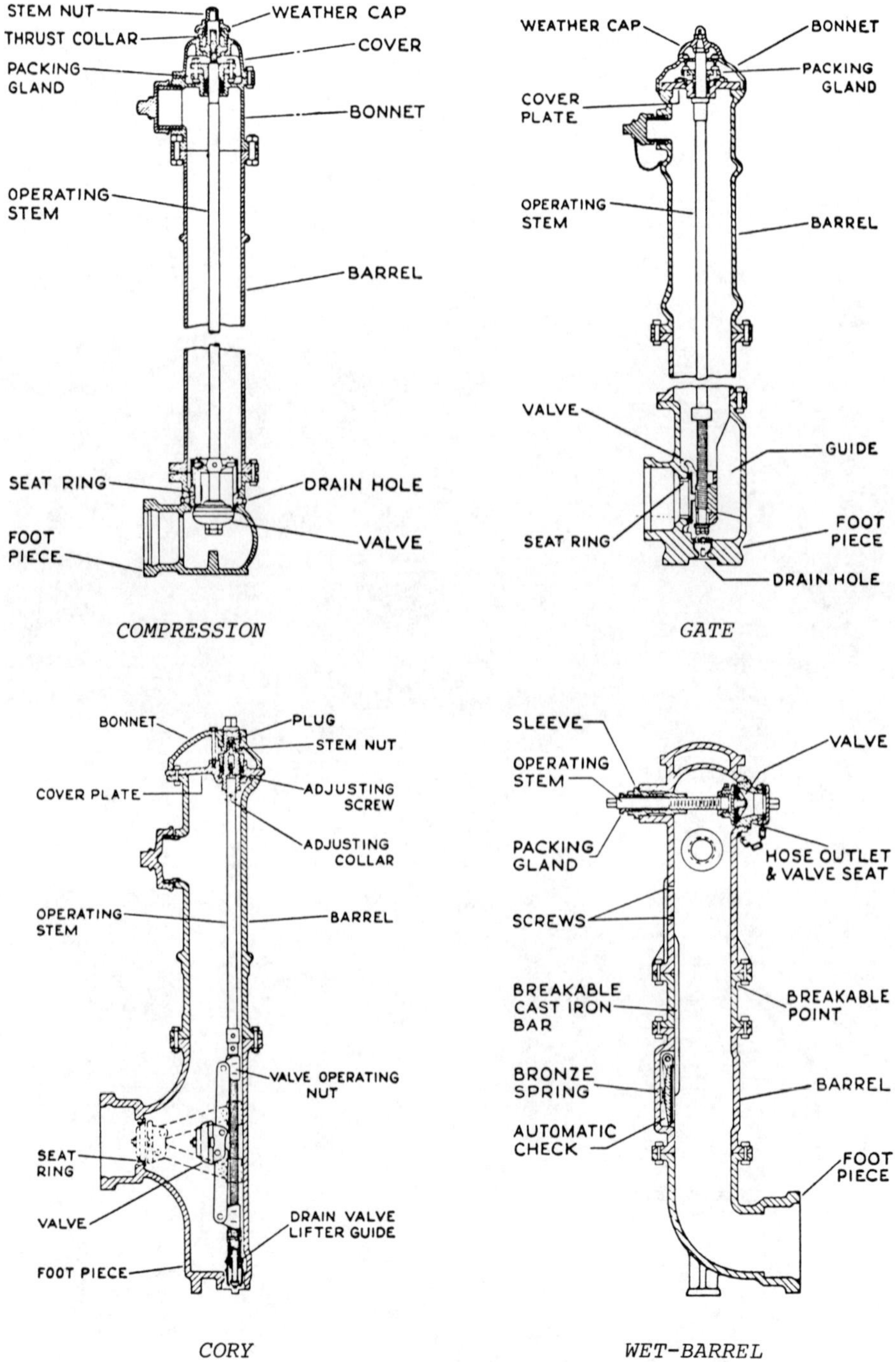

Fig. IX-7. Hydrant Types

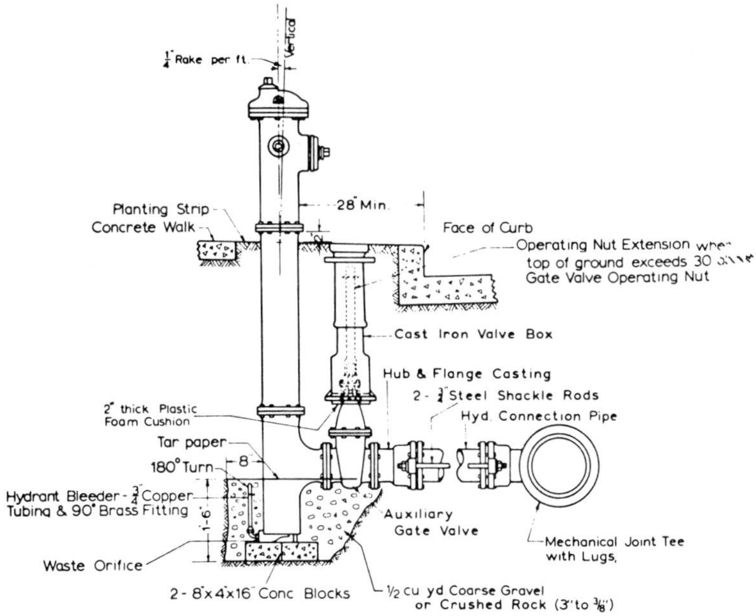

RESIDENTIAL

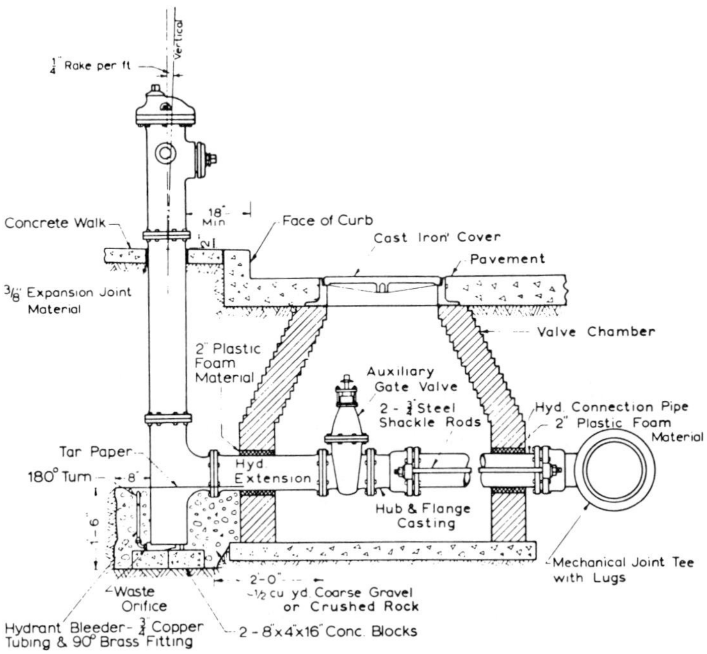

BUSINESS DISTRICT

Fig. IX-8. Typical Hydrant Settings

Valve Summary				
Valve Type	Common Size Range—*in.*	Particularly Adapted to	Main Advantages	Main Disadvantages
Gate valve	½ - 24	isolation service in pipe grids	low cost in small sizes; low friction loss; fairly good service life; ease of installation	high cost in large sizes; large valves are quite heavy; poor for throttling; should not be used where frequent operation is necessary.
Butterfly valve	3	isolation and automatic control	low cost in larger sizes for normal service pressures; some types have very short lengths, ease of operation.	higher friction loss than gate valve; must use care in selecting valve, particularly for line-break flow conditions; may cause problems when relining pipe; often leaks because of seat damage
Globe valve	½ - 24	isolation in small sizes; flow control in large sizes. pressure control	simple construction; dependable; can be used for throttling; good for pressure control; sediment or material unlikely to prevent complete closing	high friction loss; very heavy and expensive in large sizes.
Ball and plug valves	3	isolation and throttling	dependable; very low friction loss; slow shutoff characteristic minimizes closing surges; ease of operation	expensive; very heavy

so that at least one person is able to work on the valve or equipment. Care must be exercised with deep vaults to ensure the safety of persons working in the vault. Air should be pumped into the vaults, and a meter to check for oxygen and harmful gases should also be used. (Refer to Chapter XIII on safety.)

Fire Hydrants

Fire hydrants are primarily of two different designs: dry barrel and wet barrel (Fig. IX-7). Dry barrels have their operating valve at the bottom, which makes it possible to remove all the water in the hydrant. This prevents any damage that could be caused by freezing. Wet-barrel types have their operating valve at the outlet. These hydrants can only be used in areas not subject to freezing.

Post hydrants, extending above ground, are most commonly used. (Fig. IX-8 depicts such an installation.) In some locations flush-type hydrants are used. This type of hydrant has its operating nut and hose nozzles located in a cast-iron box below ground level. The box is not attached but serves as protection for the hydrant. The lid of the box has a recessed handle for easy removal.

Hydrants generally consist of four parts: the inlet connection that is attached to the water supply main, the main valve, the barrel, and the head. In some utilities a shutoff gate valve is placed so that the hydrant can be closed off shutoff gate valve to isolate them from the water main, allowing ready repair of damage of leaks.

Some barrels are designed for traffic safety. These have a breakaway point in case the hydrant is struck by a vehicle. The valve stem is also designed to break when the barrel is broken.

Four types of main valves are used. They are upward closing against a horizontal seat, downward closing against a horizontal seat, horizontal closing against a vertical seat, and a vertical gate valve. Most hydrants are designed so that it is possible to lift the valve and the stem out of the barrel by removing the head.

Hydrants are produced with different nozzle arrangements. A two-nozzle, one pumper arrangement is common. Nozzles are usually 2½ in. in diameter and pumper connections are 4½ in. in diameter. All nozzle ends should be equipped with protective caps to safeguard the nozzle threads.

Fire hydrants are often used for purposes other than fighting fires. For example, pipelines are often flushed through hydrants. It used to be common practice to use hydrants for washing streets, but this is now seldom done. Hydrants may be used to provide water for water trucks or other construction equipment. Where this occurs some type of metering arrangement should be provided. Some cities have opened hydrants in hot weather to allow children to play in the water. Personnel should be aware of unauthorized use of hydrants so that effective action may be taken.

Chapter X
Meters and Services

Introduction

This chapter is concerned with meters: different types of meters, installation procedures, maintenance, repair, and testing. The effect of meters on the system is explained. This chapter should help to explain why one type of meter is used rather than another, the problems encountered in meter installations, and the importance of meters. An employee at a water utility should learn to recognize the different types of meters and to know their operating characteristics.

Purpose of Meters

The most common use of water meters is for setting costs according to use. However, meters serve in many other ways. Main-line meters are used to measure water flow into different areas and thereby provide for system control. Meters measure the flow in and out of reservoirs and the influent to and effluent from treatment plants; they also provide information for production. Meters allow for the accurate blending of waters of different quality. Chemical dosages, such as coagulants or chlorine, can be proportioned when the water is metered. Meters also allow a water utility to determine how efficiently it is operating. If large amounts of water are unaccounted for, then the utility can begin a survey to determine where the water loss is occurring.

Meters are also important for the social and economic effects they have on the water system. Metering provides a record of water use and a fair distribution of the cost of delivering water, because customers only pay for the water they use. The periodic charge for water may encourage customers to conserve water and instill awareness of water loss through leaky faucets or pipes. An excessive water bill may be the first clue to a customer that a leak exists in his system.

Types of Meters

The most common meter is the *water service meter,* which is usually a displacement-type meter. It uses a measuring chamber of known size to determine the amount of water that passes through it. Two styles are normally encountered — the piston type and the nutating-disc type. In the piston-type meter, water flows into a chamber, thus displacing a "piston." The oscillatory or circulating (depending on design) motion of the piston is transmitted to a register that records the amount of water that has passed through the meter. The nutating disc-type meter uses a chamber containing a hard rubber disc. When water flows, the tilted disc rotates. This motion is transmitted to a register that records the volume of the water. See Fig. X-1.

Both types of displacement meters are accurate at low flows, but have a higher than normal amount of head loss due to friction.

Small amounts of sediment accumulating on the disc, piston, or chamber can cause the meter to over-register but only to a limited extent. But an excess of sediment can cause the meter to stop registering. Excessive wear in the meter normally results in under-registration.

A displacement meter should not be operated at a rate in excess of its stated capacity because this can result in excessive wear and under-registration. Displacement meters are normally used for 5/8-in—3-in. services. However, the disc type is produced in sizes up to 6 in.

Velocity, compound, or proportional meters, or meters mounted in parallel, can be used to measure large flows of water. Velocity meters, sometimes called current meters, include turbine and propeller types.

Turbine meters have a rotor mounted in a measuring chamber, or passage, which is turned by the flow of water. The rotor turns at a speed proportional to the velocity of the water flow. This movement is then transmitted through a drive train to a register. Turbine meters have little friction loss. Although turbine meters are accurate for large flows, they are inaccurate for small ones. Present-day horizontal turbine meters have improved low-end accuracy and higher maximum flows than previous instruments.

Propeller meters (Fig. X-2) are much like turbine meters in that the propeller is turned by the flow of water. This movement is then transmitted to a register. Like the turbine meter, the propeller meter is not designed for low flows or stop-and-go operation. The accurate measuring range of propeller meters is greater than that for vertical turbine meters and about the same as that for horizontal turbine meters. The propeller meter can be built within a section of pipe, or it may be saddle mounted. Friction loss with a propeller meter is less than with a turbine meter, and accuracy may also be less. Both turbine and propeller meters are most frequently used to measure sustained high flows.

Multi-jet meters are a specific class of inferential or current-type meters, as differentiated from disc or piston displacement types. In multi-jet meters the moving element takes the form of a multiblade rotor mounted on a vertical spindle within a cylindrical measuring chamber. The liquid enters the measuring chamber through several tangential orifices around the circumference and leaves the measuring chamber through another set of tangential orifices placed at a different level in the measuring chamber.

Compound meters are used where the water demand varies widely. These meters are usually a combination of two different types — displacement and turbine, for instance, as shown in Fig. X-3. The compound meter is designed so that low flows are metered by the displacement part and high flows by the turbine part. The compound meter can have one or two registers, depending upon the design. Compound meters are usually accurate over a wide range with only a slight reduction in accuracy at the transition point between the two

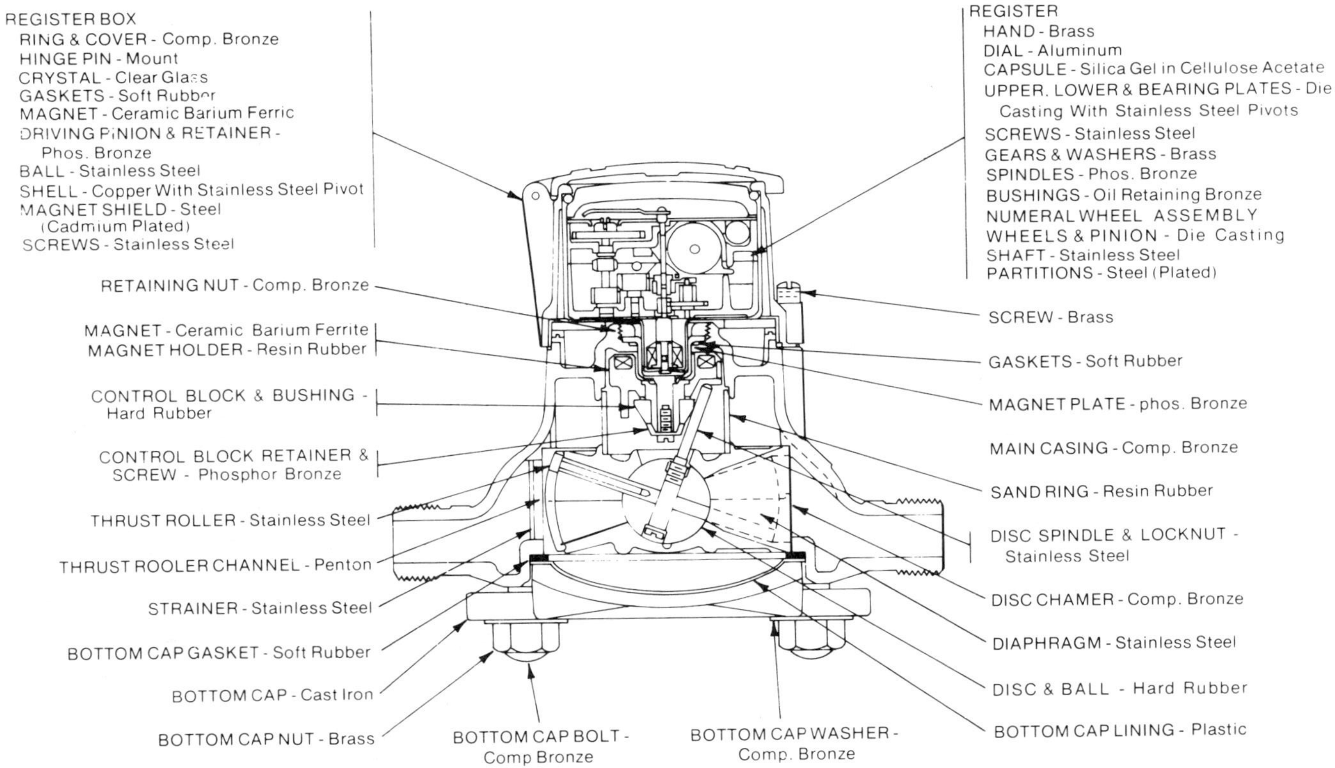

Fig. X-1. Nutating-Disk Displacement-Type Meter

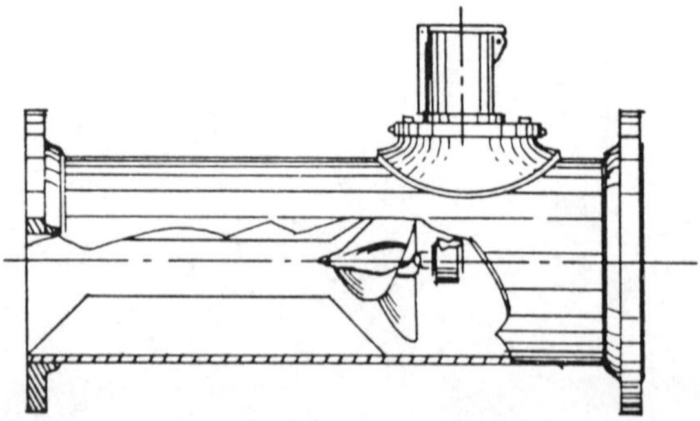

Fig. X-2. Propeller Meter

meter types within the unit. This combination is especially suitable for predominantly low or intermediate flow with only occasional high flows — conditions encountered in hotels, hospitals, schools, and office buildings.

Frequently, instead of installing a single large meter, a utility will install a series of small meters in a battery or manifold. This type of arrangement may be used to permit a utility to use displacement meters (which are the most accurate of the commercial meters), to allow for future expansion or to permit the removal of a meter without closing off the entire system. Figure X-4 illustrates such a system.

Where the metering of main lines is necessary, a meter must be used that will not interfere with the free flow of water should the meter fail. Meters that meet this requirement include venturi, proportional, orifice, pitot tube, magnetic, and sonic meters. Velocity meters of the propeller type can also be used.

A *proportional meter* uses a slightly reduced section in the pipeline to divert an amount of water into a loop such as A of Fig. X-5. Because the flow in this loop is proportional to the ratio of areas A/B a displacement or turbine meter measuring only the diverted flow can be calibrated to register the full amount of water in the main pipeline. These meters are relatively accurate but are difficult to maintain. Proportional meters, however, have little friction loss and there is little obstruction to the flow of water.

The *venturi meter* (Fig. X-6) looks much like an hourglass. The meter consists of an upstream reducer, a short throat piece, and a downstream expansion section for increasing the diameter from that of the throat to that of the downstream pipe. The amount of water passing through the meter is determined by comparing the pressure at the throat and at a point upstream from the throat. Instruments are used to record the flow rate and to totalize the flow. These instruments are normally considered as part of the meter. Venturi meters are accurate for a large range of flows and have little friction loss.

The *orifice type* of meter is often called a thin-plate orifice meter because a thin plate with a circular hole in it is installed in the pipeline between a set of flanges (Fig. X-7). As is the case with the venturi, the flow is determined by

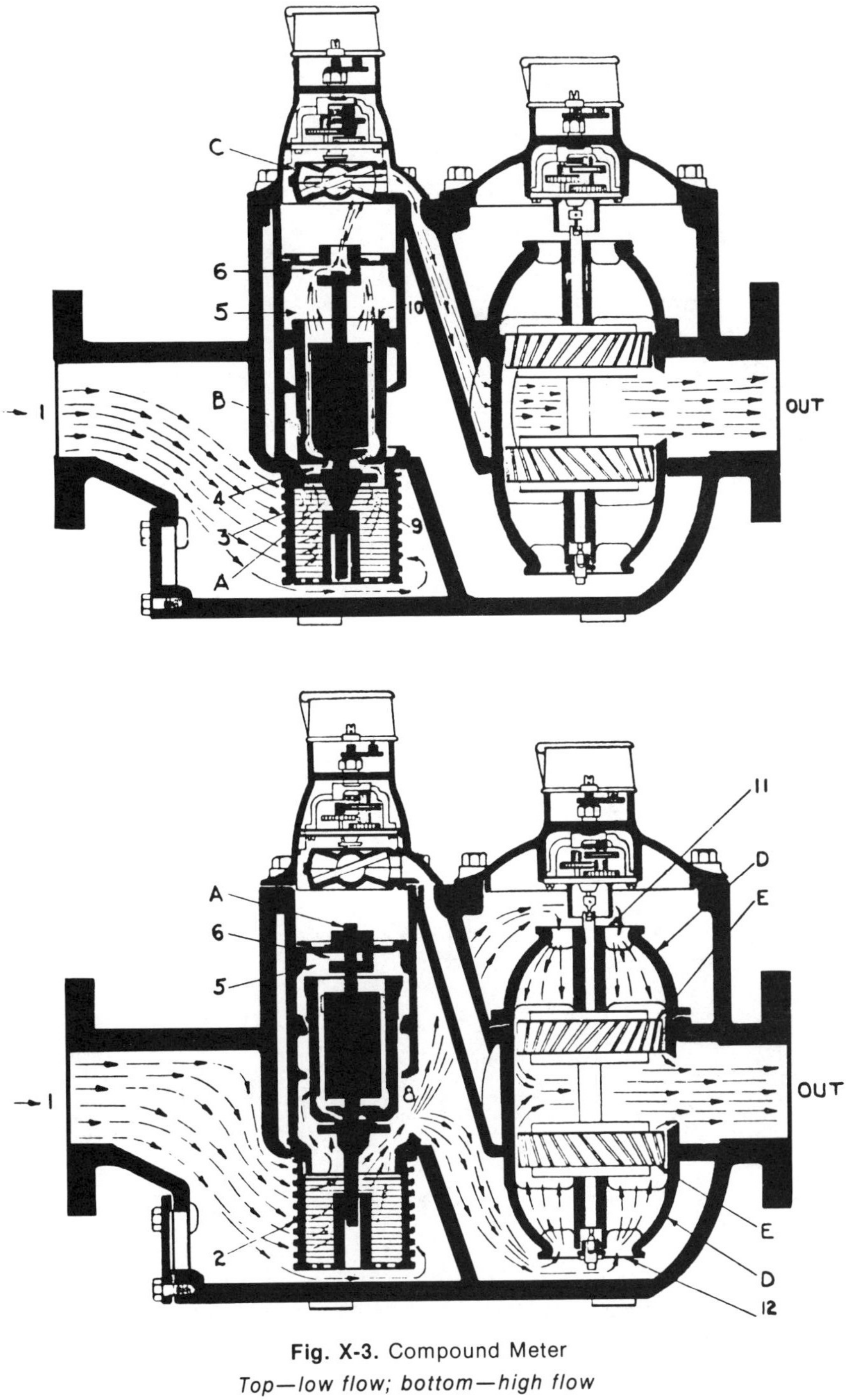

Fig. X-3. Compound Meter
Top—low flow; bottom—high flow

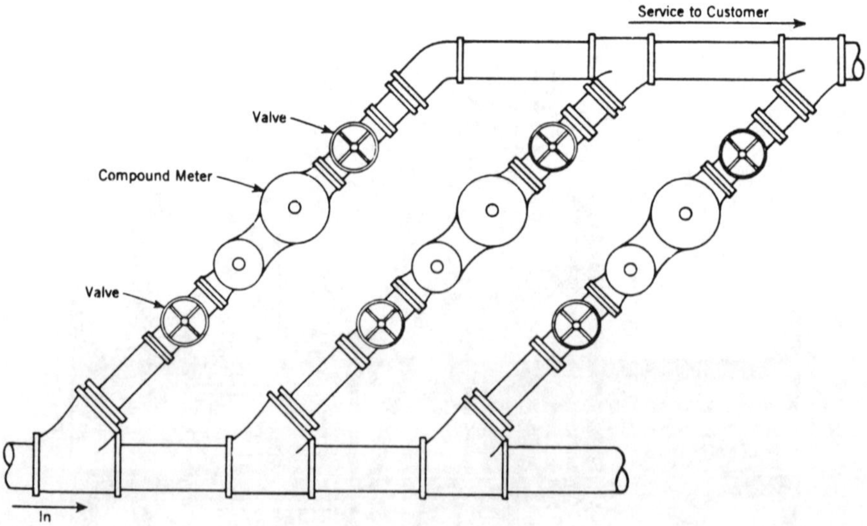

Fig. X-4. Meter Battery

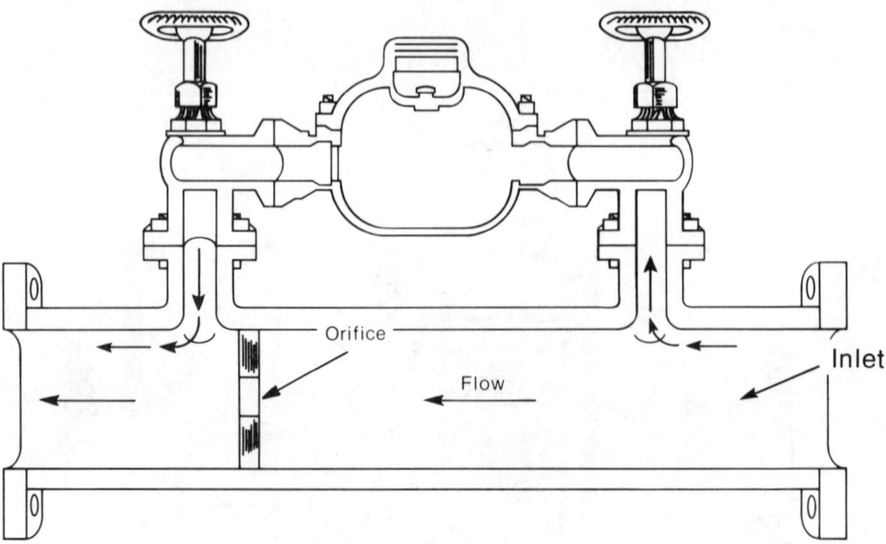

Fig. X-5. Proportional Meter

comparing the upstream line pressure with the reduced pressure at the restriction of the orifice. Pressure losses through the meter are more severe than for the venturi device; orifice meters have the advantage, however, of occupying considerably less space.

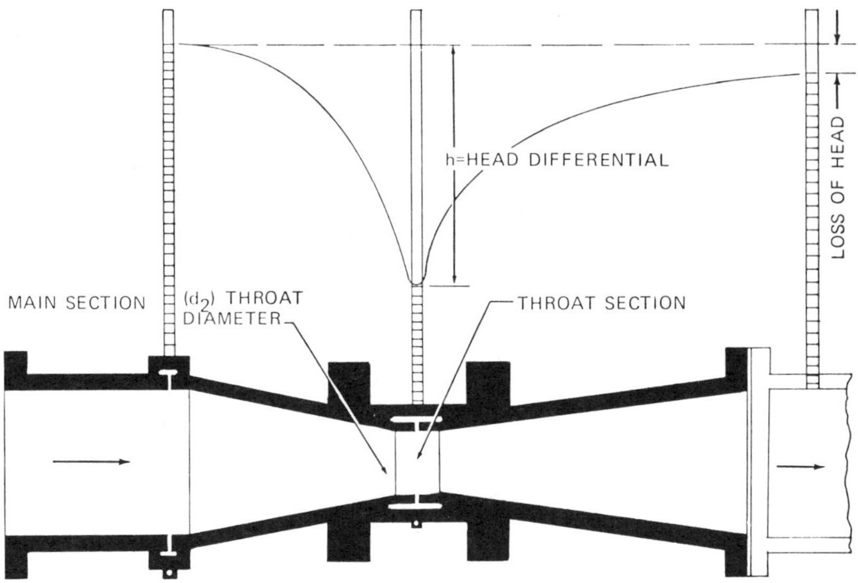

Fig. X-6. Venturi Meter

The *pitot* tube indicates flow by comparing the height difference between water in a tube with an orifice facing upstream and another with orifice pointing downstream. An indicating manometer will show greater pressure in the upstream tube because of velocity. Pitot-type meters are designed so that they can be inserted through a ¾-in. corporation stop. This type of arrangement is often used by water utilities to locate leaks and to measure the flow from fire hydrants (Fig. X-8).

Rotometers are often used in industrial plants and consist of a heavy float that rises in a tapered tube as water is forced up through it and around the float. The height of the float against a calibrated background indicates the flow. Because rotometers have a narrow range of accuracy, a battery of instruments covering different ranges is normally used.

In a *magnetic meter,* commonly called a "mag meter," water passes through a magnetic field inducing a small electric current flow proportional to the water flow. The electric current is measured and converted to a measure of water flow.

Sonic meters contain sensors (transducers) attached to the sides of a pipe. Sound pulses alternately are sent in opposite diagonal directions across the pipe. Because of the phonomenon called the "doppler effect," the frequency of the sound changes with the velocity of the water through which the sound waves are propagated. The difference between the frequency of the sound signal traveling with the flow of water and the signal against the flow of water is an accurate indication of the water flow.

Both the magnetic and the sonic types of meters are relatively new developments. These and other types of electronic meters may be used more in the future.

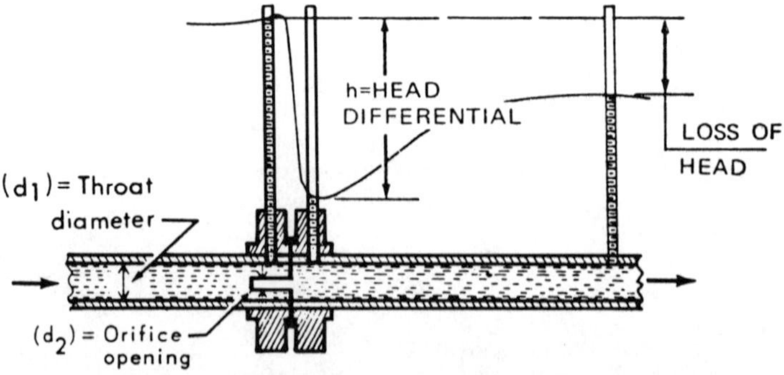

Fig. X-7. Orifice-Type Meter

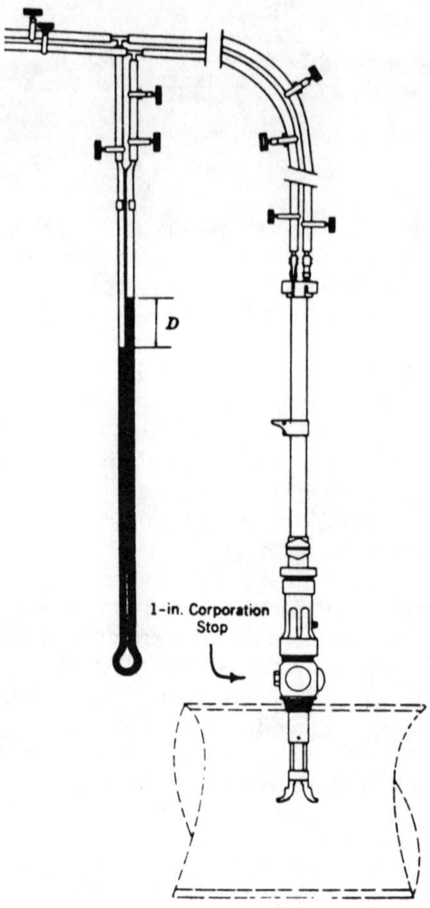

Fig. X-8. Pitot-Tube Meter

Open-Channel Flow Measurement

An open channel is any conduit through which water flows with a free surface. Examples are rivers, canals, flumes, and other uncovered conduits. Certain closed conduits such as sewers and tunnels, when not flowing under pressure are also considered as open channels.

In order to determine open-channel flows a definite relationship must be established between height of the water upstream and the channel flow. This can be accomplished by inserting a calibrated weir or Parshall flume into the channel.

Basically, a weir is an obstruction, built across an open channel, over which the liquid must pass, usually through an opening or notch in the weir plate. Most weirs are sharp crested (constructed so that the water falls free in a jet or nappe as it leaves the upstream face) and occur in three basic types — the Cipolletti weir, rectangular weir, and V-notch weir (Fig. X-9, Sketches A-C). The last usually is made with 90, 45, or 22 1/2-deg notches. Weirs are usually the simplest, least expensive devices that can be used to measure the flow of liquids in open channels.

For best accuracy the head should be at least 0.2 ft so that the overflow nappe will spring clear of the weir crest (Fig. X-9D). For small flows this generally means that a 22 1/2-deg V-notch weir is a better choice that a 90-deg V-notch weir.

If a float-operated indicator or recording instrument is to be used with the weir, it is necessary to construct a stilling well or use a scow-type float arrangement. Figure X-9E shows a typical installation for a 90-deg V-notch weir with a stilling well and recorder.

Parshall flumes are specially shaped open-channel flow sections consisting of converging throat and expansion sections, each having specific shapes and slopes. They offer significant advantages over weirs because velocity of flow within them is high. Thus, they tend to be self-cleaning, practically eliminating deposition of sediment and debris in both the flume and the upstream pool. Consequently, their accuracy is not affected by varying approach velocities nearly as much as weirs are. Additionally, they are more suited for many applications than are weirs because they can operate with very small loss in head. A disadvantage is that flume installations tend to involve more expense than weirs.

Stilling wells with floats are usually located on the outside wall adjacent to both the inlet and throat sections. A crossover pipe connects the bottom of the well to the side of the flume. Most stilling wells require cleaning at intervals to remove sediment.

A metal staff gage should be attached to the side of the flume or weir box even if water levels are obtained by automatic gages. It gives visual evidence of water level to check the accuracy of the recording instrument. After the head or depth is determined by the primary measuring device, the discharge or rate of flow can be determined by referring to the manufacturer's rating curve or table for the specified weir or flume. For a flume the zero of the staff gage must be carefully placed at the same elevation as the crest (Fig. X-10).

If proper equipment is used, open-channel flows can be measured to 95-97 per cent accuracy—acceptable for the majority of field installations.

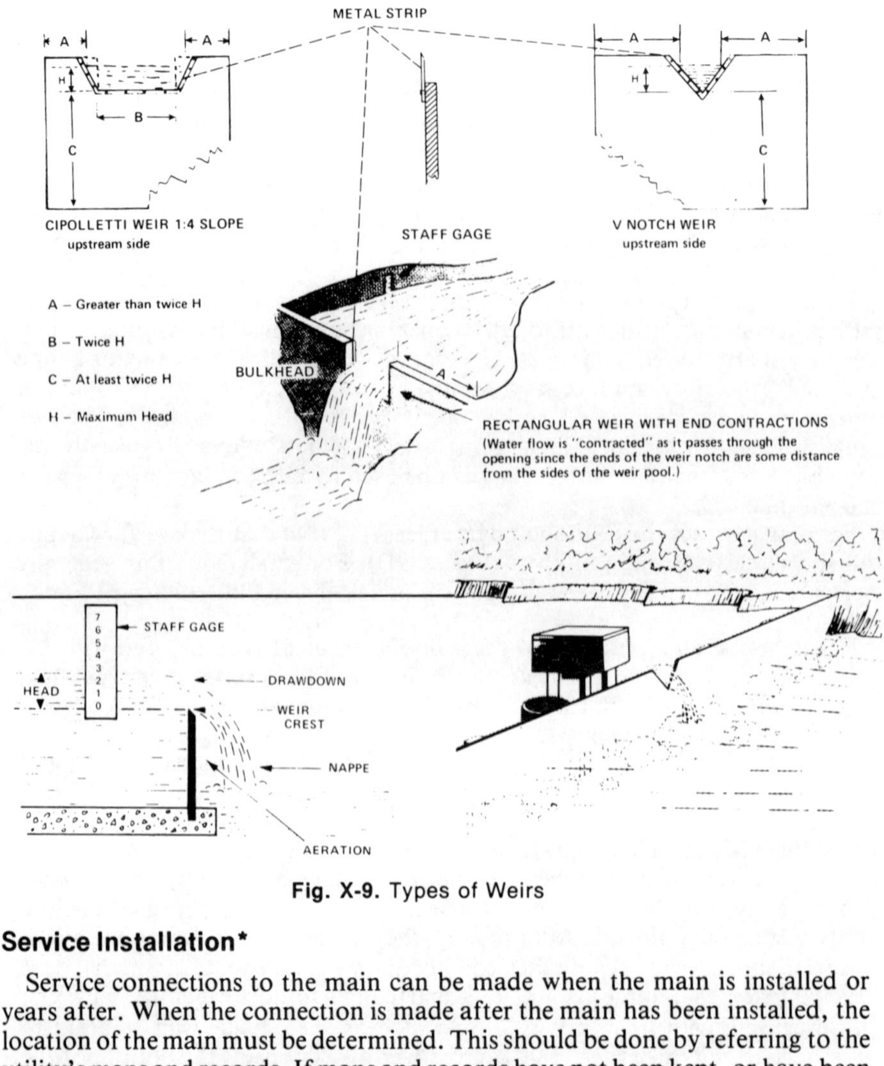

Fig. X-9. Types of Weirs

Service Installation*

Service connections to the main can be made when the main is installed or years after. When the connection is made after the main has been installed, the location of the main must be determined. This should be done by referring to the utility's maps and records. If maps and records have not been kept, or have been kept inaccurately, it may be necessary to locate the main by instruments.

The installation of the pipe for service connections is similar to the procedures covered in Chapter VIII. Utilities try to ensure that the excavation, installation, and area restoration are done in such a manner that the work will not inconvenience the public. Trenches that are excavated for service lines are usually small and should not bother local residents very much. Some utilities minimize their excavations by making use of boring and trenching machines. These devices are especially useful in developed areas with paved streets, curbs, and sidewalks.

---

*See Service Piping, Chapter VII.

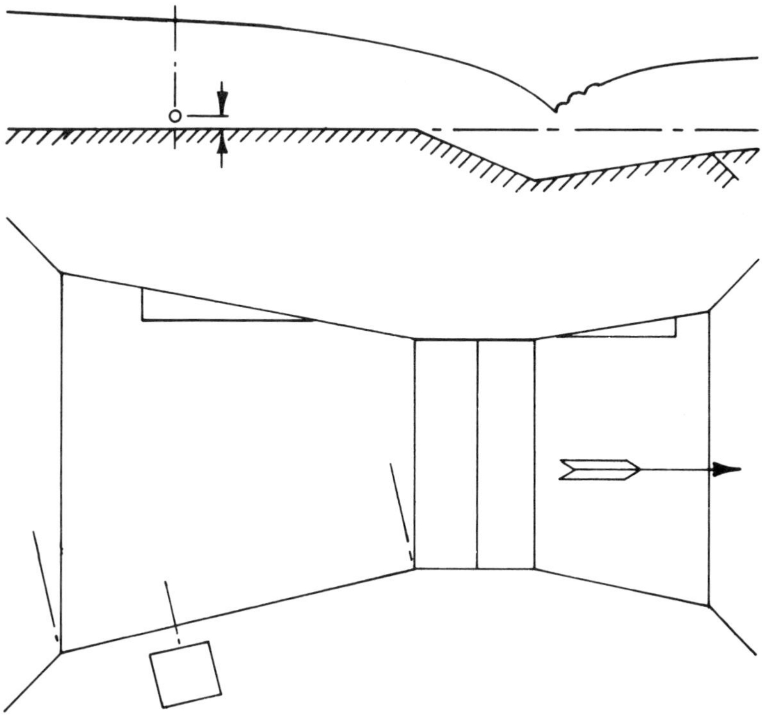

Fig. X-10. Parshall Flume

Each service connection basically consists of (1) a connection to the main (normally by installing a corporation stopcock or ferrule), (2) a length of pipe to the curb or property line, (3) a shutoff valve or curb stop (not used by some agencies), (4) a water meter, and (5) double-check valves, or a reduced pressure backflow prevention valve which should be installed between the meter and the customer's line when a potential for contamination exists. Where there is a higher degree of hazard, either an air-gap separation or an approved "reduced pressure principle" backflow prevention device should be used. The air-gap separation makes backflow impossible because of the presence of the physical separation in the supply pipe: water falls from an open pipe into a tank or receptacle. Generally, the air gap separation is at least twice the diameter of the pipe, but not less than 1 in. The reduced pressure principle backflow prevention device and double-check valve arrangements are discussed in Chapter IX.

Connections to the main can be accomplished when the pipe is empty (a dry tap) or when the main is under pressure (a wet tap). For dry taps, depending on the pipe material, the pipe can be drilled and tapped to receive the corporation stop. For wet taps, a special device enables a worker to install a stop without closing the main. When the pipe wall is too thin or soft, or the pressure is too great to permit directly tapping the pipe, a service clamp or saddle must be used. These fittings are attached around the pipe, and then the corporation stop is threaded into the fitting rather than into the pipe. These fittings can be used for both wet and dry tap into most types of pipe. Some utilities use multiple taps to provide for large service demands. This method is usually less expensive than the installation of one large tap. When multiple taps are used, they should be spaced at least 10 in. apart.

Some utilities occasionally install a large service line to serve two or three small meters, especially when the line must cross a street.

Corporation stops are available with different combinations of pipe threads for the inlet and the outlet sides of the stop. Tapered threads on the inlet side are usually desirable for reasons of strength.

Most utilities have a policy of installing stops at a specific position on the main. The most common location is usually at a point on the pipe cross section 45-deg above the horizontal centerline. However, positions 22 deg from the vertical center line and right at the horizontal centerline are also used. (Fig. X-11).

Although most service lines are made of flexible material, the line should be properly bedded to ensure that it bears evenly on the bottom of the trench. The placement of backfill material should be done carefully to ensure that the pipe is not damaged.

If a curb stop is used it can be placed in the meter box or have its own curb box. Curb boxes, like valve boxes mentioned in Chapter IX, are available in two basic designs. In one the bottom of the box is threaded so that it can be attached directly to a threaded stop. This design prevents the curb box from moving away from the stop. The other design fits over the stop. Both designs are adjustable so that the top of the box can be set at existing ground level. Where curb boxes and curb stops are used, they are usually located in an easily accessible location.

Meter Installation Practices

The choice of the size of meter installed is usually dependent on several factors. The most important is the customer's expected demand for water. Pressure considerations may also be important in determining the size of the service. Small services and meters may have too large of a pressure drop to allow for the adequate delivery of water at a suitable pressure. This problem is discussed in Chapter V, "Hydraulics."

There are seven requirements for an acceptable meter installation: (1) that the meter be watertight, (2) that it provide an upstream shut-off valve of high quality and with low pressure loss, (3) that it be positioned in a horizontal plane for optimum performance, (4) that it be reasonably accessible for service and inspection, (5) that it provide for easy reading either directly or via a remote-reading device, (6) that it be reasonably well protected against frost, mechanical damage and tampering, and (7) that it not be an obstacle or hazard to the customer or interfere with public safety.

Meters should be transported, installed, and protected with care. The size of the meter is usually described in terms of the size of the pipe for which the meter is intended. Often the meter is one size smaller than the pipe. Meters smaller

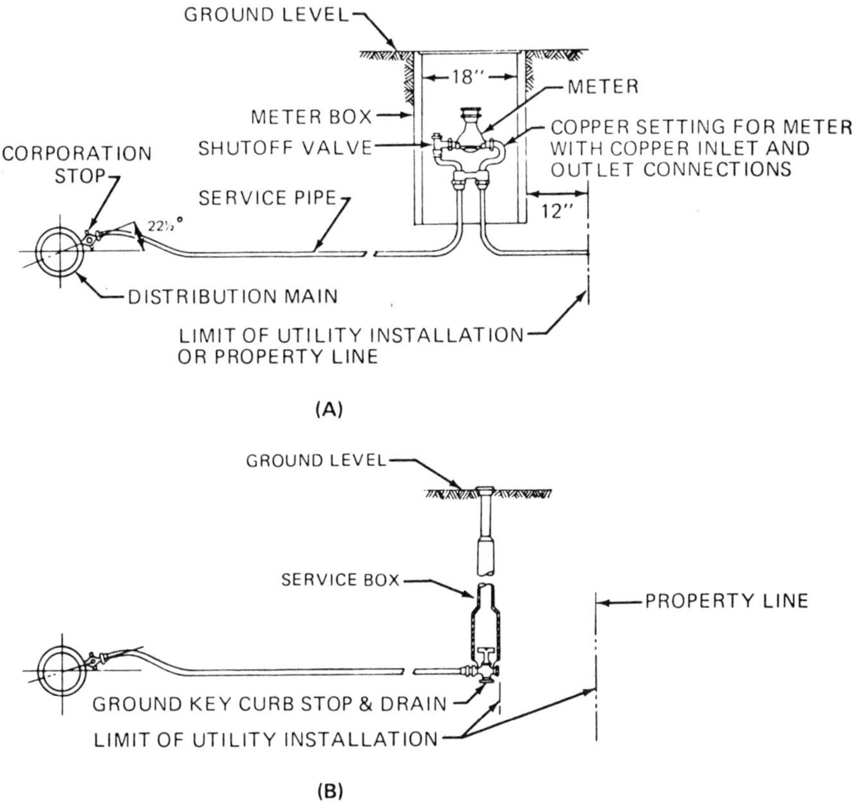

PART A SHOWS A TYPICAL SERVICE LINE WITH THE METER AT THE CURB OR CUSTOMER'S PROPERTY LINE. PART B SHOWS A TYPICAL INSTALLATION WITH A CURB BOX AT THE CURB OR PROPERTY LINE; THE METER, IF ANY, IS INSTALLED ON THE CUSTOMER'S PREMISES.

Fig. X-11. Service-Line Installation

than 1 1/2 in. usually have screw connections. Large meters are usually flanged.

Special fittings are available to make meter installations easy. These are called meter yokes and hold the stub ends of the pipe in proper alignment and spacing. Yokes also provide for a continuous electric ground (if metal pipe is used) when the meter is removed. This serves to protect meter repairmen from a possible, dangerous electric shock under certain conditions.

Meters may be installed indoors or outdoors. Where meters are installed outdoors, they are usually located in a meter box. The meter box, when installed, should have a 2-3-in. clearance above the service pipe to avoid any damage or strain should the meter box settle after installation. Large meters may require the construction of a meter pit or vault.

Indoor locations may be in basements, crawl spaces, utility rooms, or in a garage. Indoor settings must be designed for electric continuity through the service pipe (provided it is metallic) if the meter is removed. If a meter yoke is not used with such pipe, then a permanently bonded electrical grounding strap should be used (see "Ground," Chapter XII). Where dropping of water on the floor is undesirable, the installation should include a valve just downstream of the meter to prevent excessive drainage during meter service or removal.

When meters are located indoors, a remote-reading device can be installed. These devices eliminate access problems and customer's concern about strangers entering the house. These devices are usually installed in such a way that the meter can be read at a convenient outdoor location. Automatic meter-reading devices that record the reading on a punched card or tape are currently being used. These devices will probably gain more use in the future as they provide for quicker and more accurate processing.

Service Protection

Services, especially meters, should be protected from freezing. The cover over service lines must be adequate to protect against frost penetration. Because of the small volume of water in service lines, they freeze more rapidly than mains. Where outside meter settings are used, the meter pits are excavated below the frost line. Even though the meter itself is not placed below the frost line, this void, if unfilled, will reduce the chance of freezing.

Most meter manufacturers make a type of housing especially designed for service where freezing is possible (Fig. X-12). These special housings have a bottom plate designed so that, should the contents freeze, it or the holding bolts breaks before the meter is damaged. These plates are easy and inexpensive to replace.

Testing of Water Meters

Water meters are subject to wear and to deposition that may cause the meter to over- or under-register. To detect a meter that may no longer be accurate, utilities normally test every few years. How often a utility will test a meter is dependent upon the cost of the testing, the cost of the water, the chemical characteristics of the water and other factors. The AWWA has recommended that 5/8-in. service meters be tested at least once every five to ten yrs.

The three basic elements of a meter test are (1) the number of different rates of flow over the operating range of a meter, which are required to determine the overall meter efficiency, (2) the quantities of water necessary at the various test rates to provide reasonable determination of meter registration, and (3) the accuracy limits that meters must meet at different rates to be acceptable for use.

Causes of inaccuracy vary, depending upon the type of meter. Deposits on discs, propellers, housings, or in chambers may cause inaccuracies; so too may warped or broken blades or discs. Foreign material in some meters may cause them to stop registering. Also important for accurate metering is the location of the meter. Turbulence caused by poor location may cause meter readings to deviate. New meters should never be assumed to be accurate; they should be tested before being put into serivce.

The procedures for testing a single meter, for multiple meters and for the field testing of meters is given in Appendix 2 at the end of this chapter.

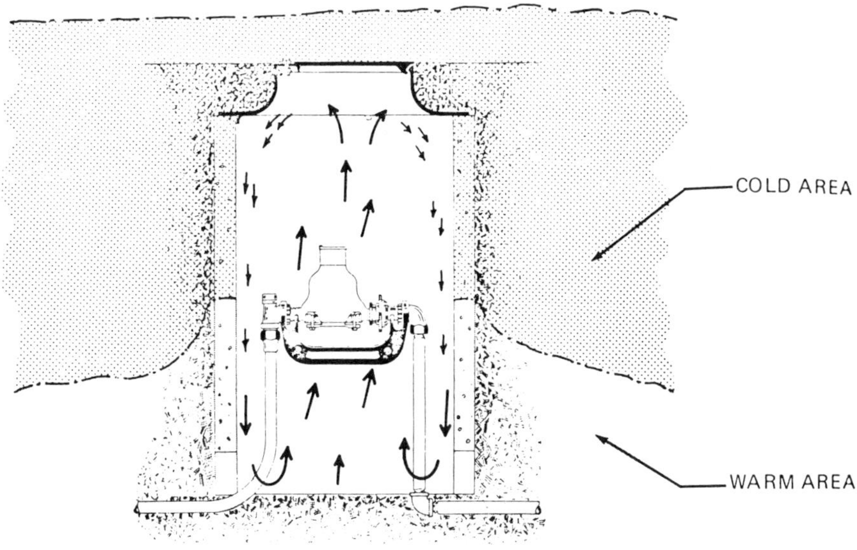

Fig. X-12. Cold-Climate Meter Installation

Maintenance and Repair

The maintenance and repair of positive displacement meters consist essentially of dismantling and thoroughly cleaning and inspecting all parts. All parts, including the measuring chamber, should be inspected for wear, pitting, swelling, distortion, and foreign material. The gear train should run freely and mesh smoothly. A meter removed from the field and not worked on for some time should be stored full of water in order to keep deposits soft and make dismantling and cleaning easier.

The following points are important to remember in meter maintenance. One should

1. Clean all meter parts thoroughly.
2. Make sure the gear train runs freely.
3. Check the action of the disc in the chamber before and after assembly in the main casing.
4. Remember that friction is just as detrimental to correct registration as slippage.
5. Store meters away from heat.
6. Use a new meter as a standard of comparison for tolerances and clearance.
7. Retest the meter for accuracy after repair.
8. If necessary, call upon the manufacturer for advice and help.

Although these principles apply for displacement meters, the maintenance and repair of other types of meters are made according to same procedures. Differences are largely the result of variations in construction.

TABLE X-1

Recommended Capacities
for
Displacement-Type Service Meters

Nominal Meter Size—*in.*	Minimum Capacity* (*gpm*)	Average Maximum Capacity† (*gpm*)	Maximum Instantaneous Capacity‡ (*gpm*)
5/8 by 3/4	¼	4	20
3/4	½	6	30
1	¾	10	50
2	2	32	160
3	4½	60	300
4	8	100	500
6	14	200	1 000

* Based on sensitivity—95 per cent registration
† Recommended maximum long-term flow rate based on 20 per cent of maximum instantaneous capacity
‡ Based on 15-psi pressure drop across 5/3- by 3/4-in. and 3/4- and 1-in. meters; 20 psi across 1 1/2-, 2-, 3-, 4-, and 6-in. meters (AWWA specification)

TABLE X-2

Recommended Capacities
for
Turbine-Type Current Meters

Nominal Meter Size—*in.*	Minimum Capacity* (*gpm*)	Average Maximum Capacity† (*gpm*)	Maximum Instantaneous Capacity‡ (*gpm*)
2	10	53	160
3	15	133	400
4	20	233	700
6	30	533	1 600
8	50	833	2 500
10	75	1 267	3 800
12	100	1 933	5 800
16	200	3 833	11 500

* Based on sensitivity—95 per cent registration
† Recommended maximum long-term rate based on 33 1/3 per cent of maximum instantaneous capacity
‡ Based on 20-psi pressure drop (AWWA specifications)
(There are turbine meters available with higher capacities for same pressure drop.)

Appendix 1
Meter Reading

Figure X-13 shows a circular register face, recording in cubic feet. To read this type of meter, start with the scale labeled with the highest number (10 000 000) and successively read the scales with lower numbers around the register until the last scale (10) is read. When the hand on a scale is between two numbers, read the lower number. When a hand is near a number, look at the next scale to see if the hand has passed the 0 point. If that hand is not on or past 0, then the lower number on the dial should be read, since the hand is actually still between the two numbers.

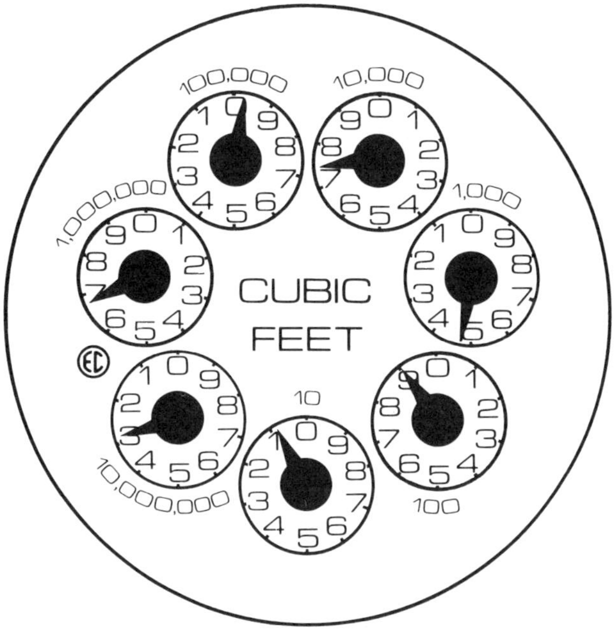

Fig. X-13. Meter Face

One complete revolution of each dial will equal the number designated by that dial. On the 10 000 000 dial, one revolution equals 10 000 000, so the numbers on the dial represent 1 000 000 through 9 000 000. On the 10s dial, the numbers correspond to 1 through 9. It is not necessary to read the dial below 10 (sometimes called the "unit" dial) if there is one, because the individual numbers are less than one, such 0.1 or 0.9. A company may choose to round off readings to the nearest 100 or 1000. The number of digits in the total meter reading will be equal to the number of zeros on the largest numbered dial label.

When reading a meter, one should note what physical units the meter is recording; usually it is gallons or cubic feet. Occasionally a meter has a multiplier on it. If so, it will be noted on the meter or the dial face by a number such "10X" or "100X." When this occurs, one must multiply the reading by the multiplier to determine the correct total. This is done simply by adding to the reading the number of zeros in the multiplier. In the illustration Fig. X-13 the largest scale has seven zeros (10 000 000) so the reading will have a maximum of seven digits. If there is a multiplier, one must add on the zeros of the multiplier. Thus, had there been a 100X multiplier, the correct reading would have a maximum of nine digits.

The most important factor in a correct reading is the proper number of digits. One should never assume that the previous reading was correct, especially for a new meter. For example, if one misreads the "2" in Fig. X-13 as a "3", the reading would be wrong by 1 million cu ft. Had the individual dials been read correctly but the reader listed eight digits rather than the correct seven, the error would be more than 24 million cu ft.

The correct reading of the meter in Fig. X-13 will be found at the end of Appendix 2 of this chapter.

Appendix 2
Meter Testing

Although the following description covers the actual steps in testing a single 5/8-in. meter, the only differences for testing larger meters are the rates of flow and test quantities used. Test equipment is essentially that shown in Fig. X-14.

Except for meters being checked upon removal from service, one should follow instructions 1-12. Testing instructions for meters removed from service are included following these itemized instructions.

1. Clamp the meter securely in the test bench, tightening only enough to make a watertight connection; there is a possibility of distorting the meter housing or extruding the washer into the water way if it is tightened too much
2. Remove the register-box cover.
3. Open the discharge valve first; then open the inlet valve gradually and run water to waste until the entrapped air is cleared. This process also ensures a full discharge line to the tank
4. Shut off the discharge valve
5. Check 100-gal or 10-cu ft tank discharge to ensure that the tank is empty and then close the tank drain valve. Tank discharge should be to an open drain so that any possible leakage of the tank drain valve can also be observed
6. Set the test hand at the zero mark by revolving the register backward beyond the zero mark and then reversing direction to bring the test dial hand forward to the mark to take up any possible backlash is the gearing (sealed registers on some magnetic-drive meters require a special test ring that sits on top of the register box and can be rotated so that the test hand starts at zero on the test dial)
7. Open the test bench discharge valve as rapidly as prudently possible to the desired rate for the maximum-flow test (15 gpm) and continue the flow at this rate until the meter test hand has made ten complete revolutions; then stop the test at the starting mark. Compare meter accuracy against the scale of the calibrated tank or, if the tank is not calibrated, weigh the water in the tank
8. Post the results of this test on the record form used
9. Refer to the change-gear chart and replace change gears with a set that should cause the meter to register within required limits if the meter being tested is a repaired meter and its accuracy does not fall within prescribed limits; repeat the test as outlined (always be sure that the meter tests correctly by rerunning the test after any change in gears is made)
10. If a repaired meter is being tested and is found to register within limits on the maximum-flow test rate, repeat steps 3-8, for the intermediate (2-gpm) and minimum (1/4-gpm) tests, discharging the test water into the 10-gal or 1-cu ft tank; the test quantity for these rates is only 10 gal or 1 cu ft and the meter register test hand completes only one full revolution
11. Remove meter from the test bench after the inlet valve of the test bench has been closed and pressure has been released by a partial opening of the discharge valve

12. Drain meter, seal register, and place dust caps on the meter spuds before the meters are put in storage

To determine the condition of a meter upon *removal from service* arrange the order of test flows according to minimum-intermediate and, finally, the maximum-flow test. If a meter has been in service for some time and is first tested at the high rate of flow, there is a possibility of freeing the disc and thereby obtaining a false impression of the meter's condition at lower rates. A meter should be tested as soon as possible after removal from service to prevent drying out of deposits in the measuring chamber since this dried condition tends to give an adverse impression of the meter's condition while it was in service.

Particular attention is called to the fact that the calibration of a tank is based on its rated capacity, and if lesser quantities are used in testing, as is sometimes done, the calibrated percentage errors are not valid. For example, a 10-gal tank may be used as a 1-gal tank by lowering the water level to the 9-gal mark and running water to the 10-gal mark. This procedure is not recommended but is sometimes carried out to save testing time. Under such conditions, each 1 per cent error on the calibration strip becomes 10 per cent.

The rate of flow indicator (rotometer) can be used as a substitute or adjunct for the calibrated tank. However, it should be noted that a rate-of-flow indicator is affected by water temperature below the 1-gpm rate. Most rate-of-flow indicators are calibrated for 70F, and viscosity changes due to temperature affect the reading.

Rate-of-flow indicators usually are equipped with a pointer to indicate true rates at the minimum flow points. Rates are raised for low-temperature water and lowered for high-temperature water. To calibrate or to use the rate-of-flow indicator, the following formula is used.

$RT = V$

where R = rate of flow in gallons per minute

T = time of flow in minutes

V = volume of test tank in gallons

Example: it takes 15 min to fill a 10-cu ft (75-gal) test tank:

$$R = V/T = 75/15 = 5 \text{ gpm}$$

This check on accuracy of low-flow calibration is required only when a marked change in temperature of the test water occurs, probably not more often than four times a year.

Multiple-Meter Testing

Multiple testing of meters is identical to the testing of one meter except for one important factor. In multiple-meter testing, it is not possible to use the meter test dials to fix upon the test quantity since each meter has a slightly different accuracy. The test quantity is discharged into the tank through several meters in series and the flow is stopped when the scale on the tank indicated the exact quantity has been delivered. Each meter's accuracy is determined from the reading of that meter's register. It is very important that the arithmetic of the test be clearly understood when meters are tested in this manner. In the tests described for testing one 5/8-in. meter, only two quantities were used: 100 gal or 10 cu ft for the maximum-flow test, and 10 gal or 1 cu ft for the intermediate and minimum-flow test. One full revolution of the test hand of a 5/8-in meter registers either 10 gal or 1 cu ft, and the test-hand circle is divided into ten equal parts. Therefore, on the maximum-flow test, when the test hand makes ten full revolutions, each subdivision of the test dial circle represents a 1 per cent variation. On the intermediate and minimum-flow tests of 10 gal or 1 cu ft, the test hand makes only one revolution, and each subdivision represents a 10 per cent variation.

When 5/8-in. meters are tested in groups and the exact test quantity is discharged into the tank, the percentage error between divisions of the test dial depends on the quantity, being 1 per cent for 100 gal (or for 10 cu ft) and 10 per cent for 10 gal (or 1 cu ft). Furthermore, whether the meter is fast or slow must be determined. If, in group testing 5/8-in. meters on a quantity of 100 gal, the test hand of one meter indicates ten full revolutions plus one additional mark (1 gal), the meter has registered 101 gal when only 100 gal were delivered, so the meter is registering 101 per cent of true quantity, or 1 per cent fast. Similarly, if the test hand of another meter has made ten revolutions minus two marks (2 gal), it has registered only 98 gal when 100 gal have been delivered. It therefore is registering 98 per cent of true quantity, or 2 per cent slow.

When 10 gal (or 1 cu ft) are used in multiple testing, each division of the test dial circle represents a 10 per cent error, and if a meter register shows on the test that the test hand has only moved 9/10 of the full circle, it is registering only 90 per cent. Unless meter registers with center sweep test hands or special test registers are used, it is necessary to interpolate between marks for more accurate readings.

Although anxiety is often expressed over the time involved in testing meters at low rates of flow and large test quantities, testing meters in multiple can reduce this concern. Furthermore, a relatively low-cost, automatic electrical cutoff valve or warning alarm can be installed in the test bench unit, which makes it unnecessary to stand by to ensure against overrun of the test.

Field Testing

Field testing is usually confined to meters larger than 2-in. in size and is recommended for all sizes of current and compound types. Very few meter shops have tank facilities for the quantities of water needed to test these large meters. Furthermore, the accuracy of some current and compound meters may be affected by the configuration of pipe fittings directly ahead of the meters.

Field testing of meters is essentially the same as shop testing, except that, instead of using a tank to measure the test water, a comparison is made between the meter to be tested and one that has been previously calibrated. The two meters are connected in series and the test water is discharged to waste. Since the calibrated meter is not 100 per cent accurate at all flows, it is necessary to use a special calibration curve to adjust for different rates of flow in order to compute the accuracy of the meter under test.

One very important point to remember in field testing is that both meters must be full of water and under positive pressure. The control valve for regula...ig flow, therefore, should always be on the discharge side of the calibrated meter. A control valve on the inlet side of the meter configuration or one located between the two meters should not be used.

The discharge of test water often presents a problem in field testing meters. An additional problem is the selection of a test meter with sufficient capacity to deliver the high rates of flow required for the maximum-flow test rate. It is often necessary to use a rate lower than those for the maximum-flow test of the large meters, and, in many instances, a maximum rate of 500 gpm or less is used. This rate, however, is usually sufficient to indicate a large meter's accuracy at the high range of flows. Rates lower than 500 gpm should be used only as an expedient and the established test rates should be employed wherever possible. It is safe to assume that the test curve will flatten out after reaching a peak registration that is approximately 10 per cent of rated capacity and then stay within required limits of registration.

The correct meter reading for the illustration on pp. X-16 is 2 697 490 cu ft.

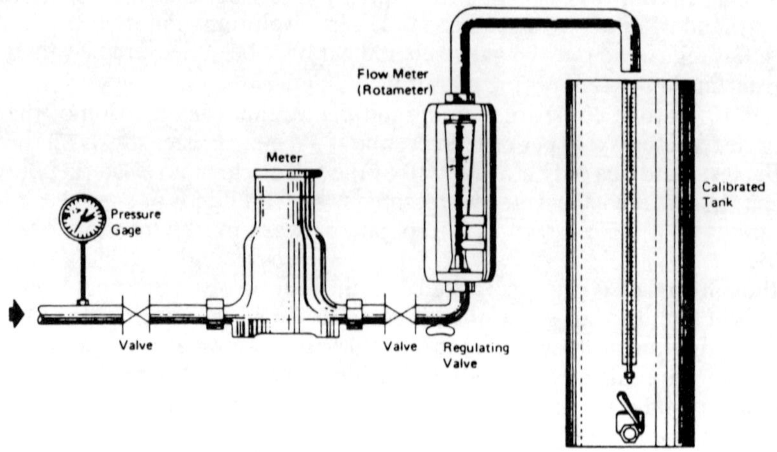

Fig. X-14. Meter Testing Installation

Chapter XI
Pumps and Motors

Introduction

This chapter is designed to introduce the reader to pumps as they are used for water distribution. The water-distribution operator should have an understanding of the different pumps used, their applications, their design differences, and the procedures necessary to maintain the pumps properly.

Types of Pumping Service

Pumps are often classified by their service application. Classifications include low lift, high lift, well, booster and standby.

Low-lift service usually involves pumping of raw untreated water from the source of supply to a treatment plant. *High-lift service* is the pumping of treated water into the distribution system under pressure. *Well service* is the pumping of water from a well. Pumps on wells may be located throughout the system. They may either pump to a central point, such as a reservoir or treatment plant, or they may pump directly into the distribution system.

Booster pumps are used for increasing the pressure in the distribution system or for supplying an elevated storage tank. They are commonly used in outlying areas of the distribution system to avoid low-pressure areas or to serve high elevations. Booster pumps are normally horizontal but may be vertical as well, as shown in Fig. XI-1.

Standby pumps, frequently called fire pumps, are pumps that are available during periods of high water demand or when pumping units are out of service for repair or maintenance. (This demand may or may not be during a fire.) These pumps may be set to come into service automatically when the demand warrants, or they may be manually operated.

Fundamental Principles

Certain general pumping terms should be understood before discussing pump operation. These terms are found in various publications, operating manuals, and sales literature for pumps. An understanding of applied hydraulics is helpful when discussing these terms, as they refer to the characteristics of water as a liquid, both at rest and moving. The relationship of the terms is shown in Figure XI-2.

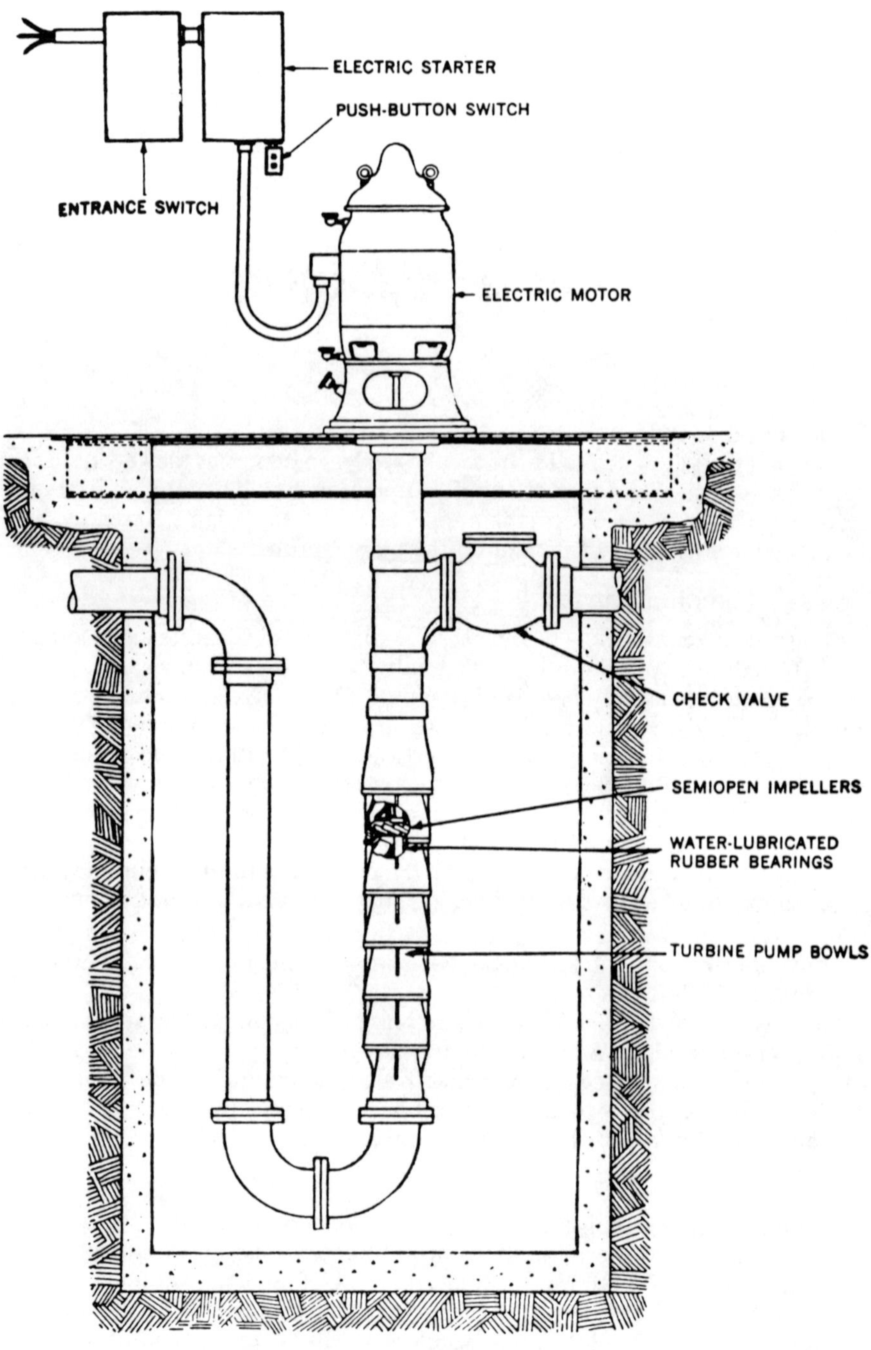

Fig. XI-1. Booster Pump

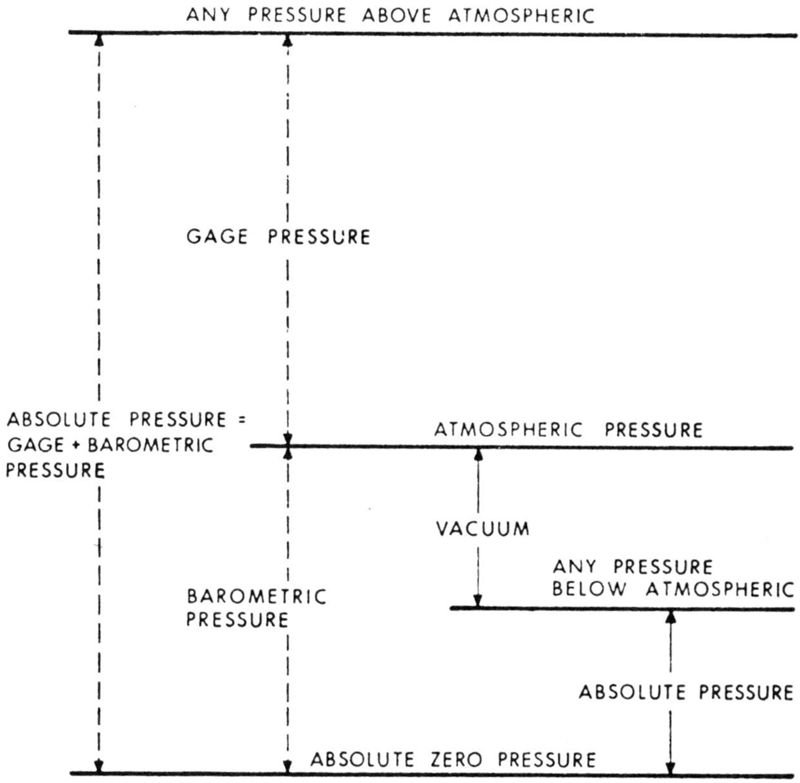

Fig. XI-2. Relationship Among Pressure Terms

Many of the terms used in describing pumping situations relate to head. Head refers to the distance (height) to which a body of water would rise if it weren't contained. Head is therefore a measure of water pressure. Head can be measured in different units; feet of liquid; pounds per square inch (psi); or inches of mercury (in. Hg) as mentioned in Chapter V. Heads are expressed in either gage or absolute units. Absolute pressure is equal to gage pressure plus the barometric air pressure (14.7 psi at sea level).

The vertical distance from the liquid supply level to the pump centerline, the pump being above the supply level, is the *static suction lift.* The vertical distance from the liquid supply level to the pump centerline, the pump being below the supply level, is the *static suction head.* The vertical distance from the pump centerline to the point of free delivery of the liquid is the *static discharge head.* Care must be exercised to see that the point of free delivery is used when computing static discharge head. This point may be difficult to determine. *Total static head* is the vertical distance between the supply level and the discharge level of the liquid being handled (see Fig. XI-3)—that is, the distance between the supply level and the point of highest lift.

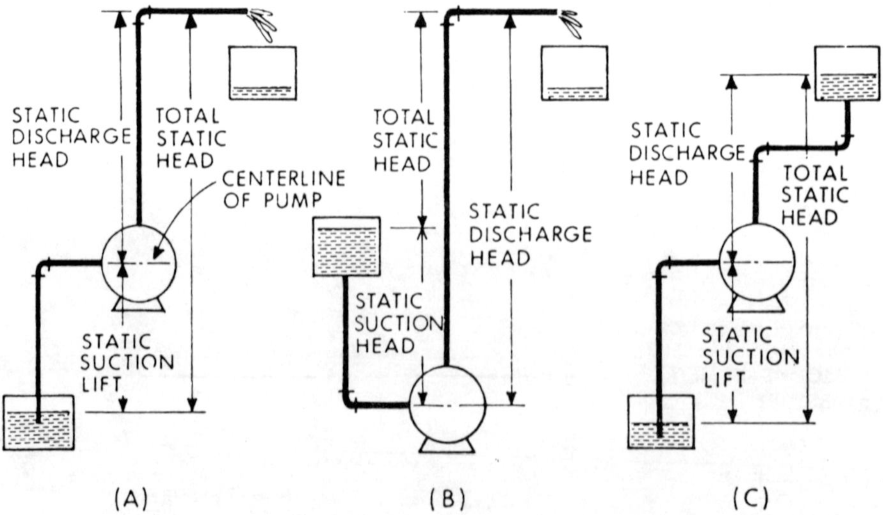

Fig. XI-3. Head Terms Used in Pumping

Friction head is the equivalent head needed to overcome the resistance of the pipe, valves, and fittings in the pumping system. It exists on both the suction and discharge sides of the pump. Factors that affect the friction head are the flow rate, pipe size, interior condition of the pipe and the quality of the water. *Entrance and exit losses* are head losses that occur at the entrance and exit of the piping and at abrupt changes in piping sizes.

Suction lift is the sum of the static suction lift, the suction friction head and the entrance loss in the suction pipe. The suction friction head includes the friction in the pipe and all fittings in the suction line. *Suction head* is the static suction head minus the suction friction head and suction piping entrance loss. *Discharge head* is the sum of the static discharge head, the discharge friction head, and the discharge *velocity head;* that is, the pressure attributable to the kinetic energy of the moving water. *Total head* is the sum of the suction lift and the discharge head, or, where there is a suction head, the difference between the discharge and suction heads.

Pump efficiency is a measure of a pump's ability to convert motor brake horsepower to water horsepower. One horsepower equals the lifting of 33 000 lb one foot in one minute. If one knows the water flow and the total pump head, the total weight of the water and the distance it is to be raised in a given time the water horsepower can be computed. Because of friction losses, impeller slippage, and other factors, the converted water horsepower must always be less than the brake horsepower. Pump efficiency, normally expressed as a percentage, changes as the operating conditions change.

Every pump has a capacity rating stated in either gallons per minute (gpm) or million gallons per day (mgd) at a certain head. However, this capacity rating is dependent upon operating conditions. If the suction or discharge condition or the speed of the motor is varied, then the capacity will vary.

In order to select the proper pump for the job, pump performance characteristics curves are drawn up. A typical characteristic curve diagram is shown in Fig. XI-4. All manufacturers supply performance characteristics for their pumps; these indicate how the pump capacity changes with variations in discharge pressure or suction conditions.

Several different types of pumps are used by water utilities. The most commonly used is the centrifugal pump; however, positive displacement pumps, and jet and hydraulic rams are also used.

Centrifugal Pumps

The advantages of centrifugal pumps are: (1) low first cost, (2) compactness, (3) absence of valves and pistons, (4) low maintenance cost and low rate of depreciation, (5) smooth flow and uniform pressures, (6) simplicity of design and ease of operation repair, (7) freedom from shock injury, (8) high rotating speed with direct connection to turbines or motors possible, (9) ability to handle "dirty" water, (10) increased output with line pressure drops, and vice versa,

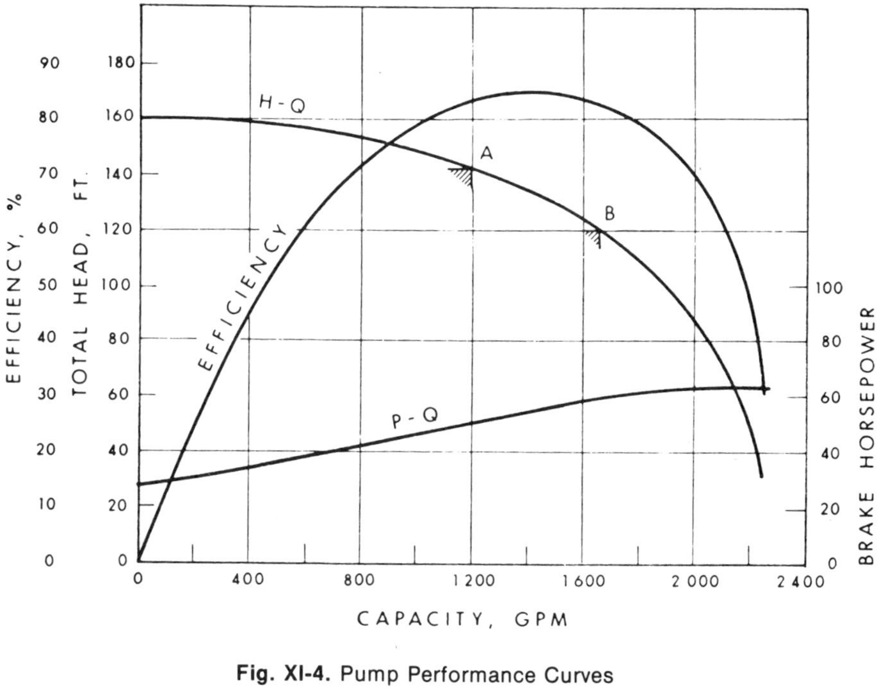

Fig. XI-4. Pump Performance Curves

(11) low starting torque, and (12) adaptability of automatic and semiautomatic control.

However there are disadvantages. Centrifugal pumps with impellers above the water source have a maximum theoretical suction lift of 34 feet at sea level and in practice, 15 - 25 ft is considered maximum. The actual suction lift of the pump is dependent upon the characteristics of the impeller, the speed, and the capacity of the pump.

Centrifugal pumps have a rotating impeller within the pump case (Fig. XI-5).

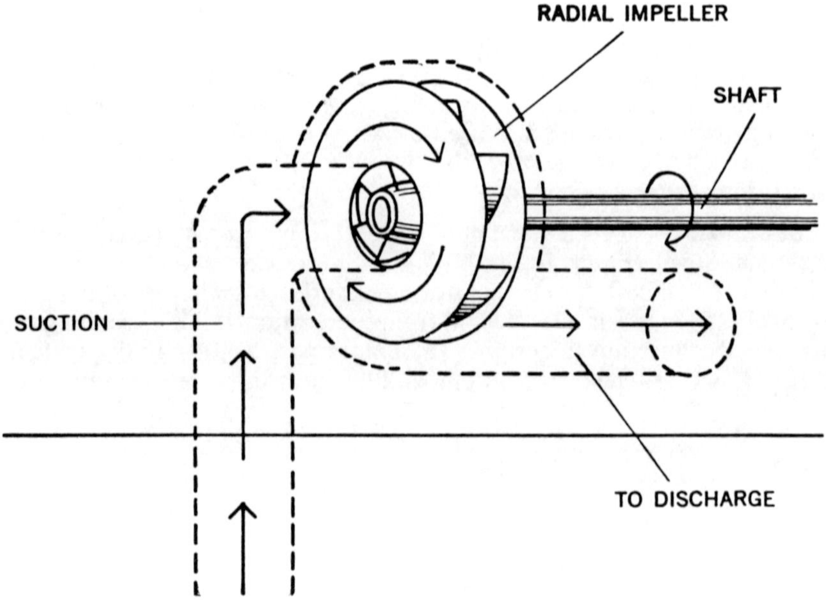

Fig. XI-5. Radial-Flow Pump

Water enters at the center of the impeller. When the impeller is rotated, water in the pump is forced radially out of the pump because of centrifugal force. This causes a vacuum condition at the eye or center, which provides the necessary condition to lift or move the water. Water is continuously being drawn toward the vacuum and at the same time being discharged by the centrifugal force of the impeller. This action provides for a smooth and continuous flow of water.

Although all centrifugal pumps operate on the same principles, they come in many different designs, arrangements, and types.

The most commonly used centrifugal pump is the *volute design*. In the volute pump the impeller discharges into a progressively expanding spiral casing. The proportions of the casing are designed to produce equal velocity of the liquid all around the casing's circumference. With this design the velocity of the liquid is gradually reduced as it flows from the impeller to the discharge. This results in the energy of the pump, which at this point is a velocity force or head, being converted to a form of energy that is static pressure force or head.

Another common centrifugal pump is the *diffuser type.* Diffuser centrifugal pumps, which are often called turbine pumps, operate similarly to the volute type. However, instead of having the case designed to convert the energy from velocity to static pressure, the diffuser type has specially designed vanes on the impeller, which perform the same task. The diffuser type is somewhat more efficient than the volute type in converting velocity pressure to static pressure.

Multistage centrifugal pumps (Fig. XI-6) are often used where a single-stage centrifugal pump will not efficiently provide the necessary discharge head: two or more impeller stages are connected together in a series. The impellers of the pumps all operate on the same shaft. The discharge of the first stage is connected to the suction of the next and so on. As the same water passes through each stage, the flow remains the same as for a single stage, but each stage increases the discharge head. Thus, it is possible to pump water to heights for which a single-stage centrifugal pump could not be efficiently designed. (Regardless of staging, maximum theoretical suction lift of any pump is about 34 ft.)

Centrifugal pumps may have a single suction inlet or a double suction inlet. This means that the water is admitted on one side of the impeller if it is a single suction and both sides if it is a double suction.

Turbine pumps, as illustrated in Fig. XI-7, which should not be confused with the diffuser-type, are considered a type of centrifugal pump. This pump does not rely solely upon centrifugal action to do the necessary pumping. In the turbine pumps the water does not flow freely from the impeller vanes but is held in the pump until the water has rotated about the pump. The water, when released at the discharge, has a higher pressure than the common centrifugal pump. These can be furnished with multiple stages for higher discharge pressures.

Mixed-flow pumps (Fig. XI-8) are a type of centrifugal pump that produces

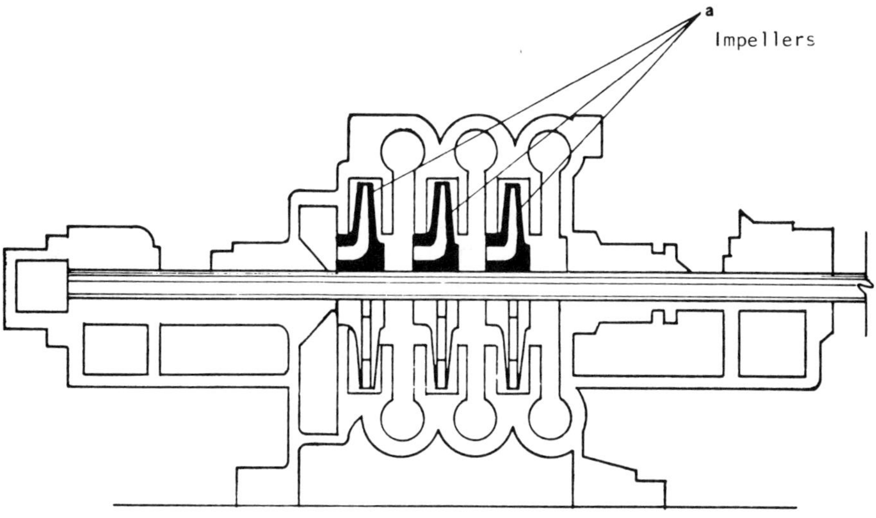

Fig. XI-6. Multistage Centrifugal Pump

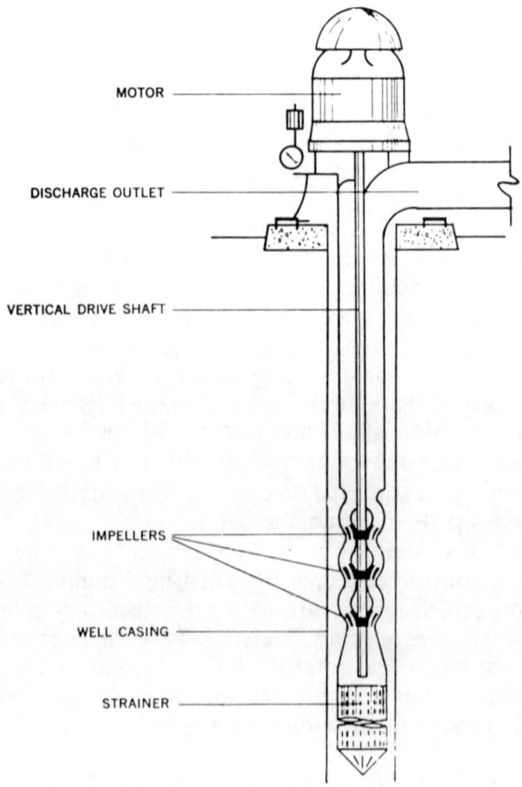

Fig. XI-7. Turbine Type Centrifugal Pump

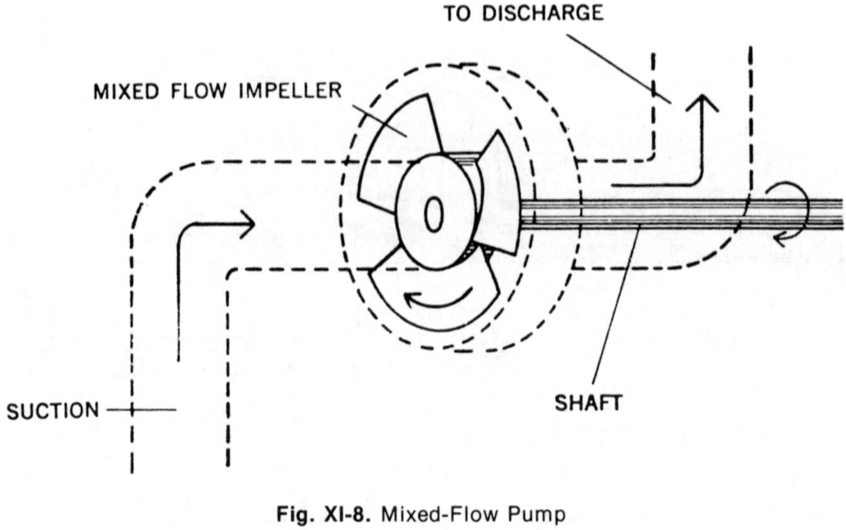

Fig. XI-8. Mixed-Flow Pump

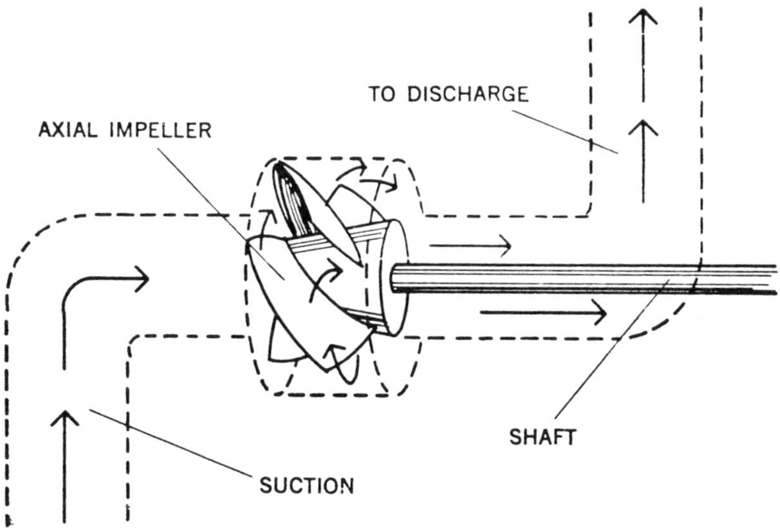

Fig. XI-9. Axial-Flow or Propeller Pump

the necessary water pressure by a combination of centrifugal force and the lift provided by the vanes of the impeller. [Axial-flow or propeller pumps (Fig. XI-9) may or may not be considered centrifugal pumps since they rely almost solely upon the lift of the impeller vanes to produce the flow of water.]

Submersible (deep well) pumps are multistage vertical turbine pumps constructed so the pump and motor can be submerged; they require no suction lift. They are completely out of sight and quiet and can be installed directly in the well casing or pipeline. Regardless of design, they can force water to considerable elevations.

Mechanical Details of Centrifugal Pumps

Impeller size may vary greatly from one pump to another, depending on the operating head and discharge conditions they are designed to handle. Basically, however, the principle of operation is the same. Water enters the impeller along the pump shaft where it is picked up by curved vanes, which efficiently change the flow direction from axial to radial. Both pressure and velocity increase as the water is impelled outward and discharged into the pump casing. The overall impeller diameter, inlet area,vane curvature, and speed affect impeller performance and can be modified by the manufacturer to attain required operating characteristics.

Almost all impellers for water-utility use are made of bronze, although most manufacturers offer cast iron or stainless steel as alternates.

The Case: As the water leaves the impeller it is traveling at high velocity in both radial and circular directions. To minimize energy losses due to turbulence and friction, the casing is designed to convert the velocity energy to additional pressure energy as smoothly as possible. This is accomplished in most every water-utility pump by casting the case in the form of a smooth volute, or spiral, around the impeller outside diameter.

Cases are available in a variety of forms—single or double suction, side, bottom, or end suction, side or bottom discharge, and multistage. Cast iron is the most common construction material, but ductile iron, bronze, and steel are usually available on special order.

Shaft: The impeller is rotated by a pump shaft, usually machined of steel to provide proper strength and stiffness. A common method of securing the impeller to the shaft is with a key and light shrink fit. This method offers simplicity of design with reasonable ease of maintenance. In end-suction design the impeller is mounted on the end of the shaft and held in place by a nut. There are several other methods of mounting impellers, but these are the most common in current use.

Packing glands: Since water enters the impeller eye at the shaft - impeller junction, the point where the shaft protrudes through the case is subjected to suction-lift or suction-head pressure. If the pump is operated under a suction lift, the vacuum at the impeller inlet will draw air into the water through the clearance between the case and shaft. If the pump is operated with a suction head, impeller inlet pressure will be above atmospheric pressure, and water can discharge along the shaft. Thus, it is necessary to have a packing gland or seal wherever the pump shaft protrudes through the case.

Packing glands usually consist of several separate rings of graphite-impregnated asbestos around the shaft. The gland and case are designed so that the rings can be compressed to bear against the shaft and thus stop air from leaking in or water from leaking out. Asbestos material is used because heat is generated by even the light friction of the packing rings against the pump shaft. The asbestos is impregnated with graphite or polytrifluoroethlyene to provide a small measure of lubrication. It is important that packing glands be adjusted properly. (This will be discusssed in more detail later.)

If the pump must operate under a high suction head (60 psig or more) the suction pressure itself will act to compress the packing rings regardless of the operator's care, and packing replacement will usually be excessive. Most manufacturers will recommend use of a mechanical seal under these conditions. The seal in this case is made between two ground and polished surfaces: one is attached to and rotates with the shaft; the other is attached to the case. Contact between the seal surfaces is maintained by a spring load. The advantage of a mechanical seal is that it can be hydraulically balanced so that the wearing force between the ground surfaces does not vary regardless of the suction head. In addition, there is little or no leakage from a mechanical seal. The major disadvantages of mechanical seals are the relative difficulty of replacement compared with packing rings and the fact that seal failure is usually accompanied by excessive leakage. Packing glands normally wear gradually and the wear in most cases can be detected long before leakage becomes objectionable.

Lantern ring: When a pump operates under suction lift, pressure at the impeller inlet is below atmospheric, and air will enter the water stream along the shaft if the packing does not provide an effective seal. Even with graphite and asbestos it may be impossible to tighten the packing sufficiently without causing excessive heat and wear on the packing and shaft or shaft sleeve. To prevent this,

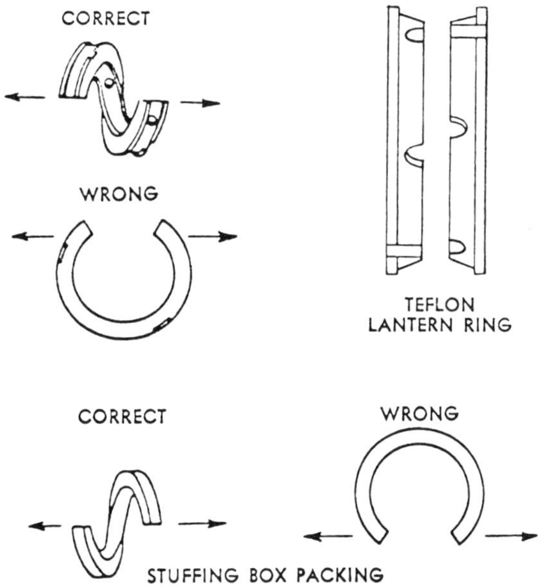

Fig. XI-10. Packing and Its Placement

a lantern ring (Fig. XI-10) is inserted in place of one of the center packing rings. The ring has radial holes through which pump discharge water is fed and then drawn into the suction stream through the packing rings. This seal water prevents air from entering the water stream and provides an additional lubricant for the packing.

Shaft sleeves: Most manufacturers provide replaceable sleeves for the packing rings to bear against. If sleeves were not used, long operation would eventually cause the shaft to wear and require replacement. A shaft could also be ruined almost instantaneously by overtightening of the packing gland. The use of shaft sleeves requires only that the sleeves themselves be replaced. In addition, the sleeves are usually made of bronze alloy, which is much more resistant to the corrosive effects of water than is steel. Stainless-steel sleeves are also usually available if the water contains abrasive elements.

Wear rings: In all centrifugal pumps a flow restriction must exist between the impeller discharge and suction to prevent excessive circulation of water between the two. This restriction is accomplished by wear rings, one of which is mounted in the case and the other on the impeller. The rotating impeller ring and the stationary case ring are machined so that the running clearance between the two effectively restricts leakage from impeller discharge to suction. Inexpensive pumps do not have separate wear rings as such, but the case and impeller are machined to prevent this leakage. Wear will occur on both the impeller and case restrictive surfaces when leakage of water occurs. Without replaceable rings this wear will cause a significant decrease in pump efficiency. Most manufacturers produce pumps with replaceable impeller and/or case rings. Rings are normally machined from bronze or cast iron, but stainless-steel rings are available.

Bearings: Most modern pumps are equipped with ball-type radial and thrust bearings. These bearings are offered with either grease or oil lubrication and provide good service in most water-utility applications. They are reasonably easy to maintain and readily available if replacement is required. A ball bearing will usually give audible notice of impending failure so that time can be provided for a planned shutdown.

Operation

In starting an electric-motor-driven centrifugal pump, the following rules of procedure should be observed:

1. Check the lubrication.
2. Prime the pump and make sure that the pump and suction piping are free of air.
3. Reduce electric current and water surge on startup by starting the pump with the discharge valve closed or throttled.
4. Open the discharge valve slowly as soon as the pump is running.
5. Check the packing glands and see that water seals are properly functioning.
6. Check the running amperage and investigate any abnormal demands.

Pumps, when operating continuously, should be checked on a routine basis. Normally checks include observing and recording pump pressures and output, and the pump's current demands. Visual checks should be made for excessive or abnormal noise, vibration, heat, and odor. If the packing glands are leaking excessively they should be tightened so that there is only a small amount of leakage but no increase in heating. Care should be taken that the packing is not tightened excessively since that could cause damage to the pump.

There are a number of causes of centrifugal pump trouble which one would be well to know about. These are listed in the centrifugal pumps trouble shooting chart shown in Table XI-1.

Priming: A centrifugal pump cannot operate unless the impeller is submerged in water. Because of this a variety of methods exists to "prime" the pump; that is, to fill the case so that the pump may begin developing a suction at the eye of the pump. One such means is shown in Fig. XI-11. Another method commonly used is to provide the pump with a foot valve that is designed to keep the pump and the suction line full of water. However, this method does not always work as the foot valve may leak, causing the pump to lose its prime. Pumps may also have a priming tank or auxiliary water supply whereby water may be bled into the pump casing and the air discharged. An electric or hand-operated vacuum pump may be provided whereby the air may be pumped from the pump casing causing water to flow into the suction inlet and the pump casing. Priming can be avoided altogether if the pump can be located below the water level so that there is pressure on the suction side.

Positive-Displacement Pumps

Positive-displacement pumps were at one time the most commonly used pumps in the water-utility field. However, the centrifugal pump's design advances and advantages of operation have made the positive-displacement type pump fall into disuse. This type of pump is now used primarily for special situations because it must produce a pulsating flow rather than a steady one.

Table XI-1

Centrifugal Pump Trouble Shooting Chart

Problem	Possible Causes
Pump does not deliver water	1-4, 6, 11, 14, 16, 17, 22, 23
Insufficient capacity delivered	2-11, 14, 17, 20, 22, 23, 29-31
Insufficient pressure developed	5, 14, 16, 17, 20, 22, 29-31
Pump loses prime after starting	2, 3, 5-8, 11-13
Pumps require too much power	15-20, 23, 24, 26, 27, 29, 33, 34, 37
Stuffing box leaks excessively	13, 24, 26, 32-36, 38-40
Packing has short life	12, 13, 24, 26, 28, 32-40
Pump vibrates or is noisy	2-4, 9-11, 21, 23-28, 30, 35, 36, 41-47
Bearings have short life	24, 26-28, 35, 36, 41-47
Pump overheats and seizes	1, 4, 21, 22, 24, 27, 28, 35, 36, 41

Key to Possible Causes of Operation Difficulties with Centrifugal Pumps

1. Pump not primed
2. Pump or suction pipe not completely filled
3. Suction lift too high
4. Insufficient margin between suction pressure and vapor pressure
5. Excessive amount of air or gas in liquid
6. Air pocket in suction line
7. Air leaks into suction line
8. Air leaks into pump through stuffing boxes
9. Foot valve too small
10. Foot valve partially clogged
11. Inlet of suction pipe insufficiently submerged
12. Water-seal pipe plugged
13. Seal cage improperly located in stuffing box, preventing sealing fluid from entering space to form seal
14. Speed too low
15. Speed too high
16. Wrong direction of rotation
17. Total head of system higher than design head of pump
18. Total head of system lower than pump design head
19. Specific gravity of liquid different from design
20. Viscosity of liquid differs from that for which designed
21. Operation at very low capacity
22. Parallel operation of pumps unsuitable for such operation
23. Foreign matter in impeller
24. Misalignment
25. Foundations not rigid
26. Shaft bent
27. Rotating part rubbing on stationary part
28. Bearings worn
29. Wear rings worn
30. Impeller damaged
31. Casing gasket defective, permitting internal leakage
32. Shaft or shaft sleeves worn or scored at packing
33. Packing improperly installed
34. Incorrect type of packing for operating conditions
35. Shaft running off center because of worn bearings or misalignment
36. Rotor out of balance, resulting in vibration

37. Gland too tight, resulting in no flow of liquid to lubricate packing
38. Failure to provide cooling liquid to water-cooled stuffing boxes
39. Excessive clearance at bottom of stuffing box between shaft and casing, causing packing to be forced into pump interior
40. Dirt or grit in sealing liquid, leading to scoring of shaft or shaft sleeve
41. Excessive thrust caused by mechanical failure inside pump or by failure of hydraulic balancing device, if any
42. Excessive grease or oil in antifriction bearing housing or lack of cooling, causing excessive bearing temperature
43. Lack of lubrication
44. Improper installation of antifriction bearings (damage during assembly, incorrect assembly of stacked bearings, use of unmatched bearings as a pair)
45. Dirt getting into bearings
46. Rusting of bearings due to water in housing
47. Excessive cooling of water-cooled bearing, resulting in condensation in bearing housing of moisture from atmosphere

---

Also the volume produced is ordinarily too small. The pump is expensive to purchase and maintain, and it should not be operated against a closed discharge because excessive pressure buildup occurs. The positive-displacement pump is still occasionally used in small systems under special circumstances. Uses include the "mud sucker" diaphragm pump used to dewater excavations, metering pumps, priming pumps, and special-purpose pumps requiring high pressure with low volume. Air pumps (compressors) are likely to be positive-displacement pumps. Because of the possibility of encountering a positive-displacement-type pump, personnel in the water-distribution field should be aware of its operating aspects.

Positive-displacement pumps are normally divided into reciprocating and rotary types. *Reciprocating pumps* (see Fig. XI-12) include piston, plunger, and diaphragm types.

Piston pumps operate by creating a vacuum in a cylinder by the withdrawal of the piston—much like in an automobile's engine where the gas and air mixture is drawn into the cylinder by the downward movement of the piston. After entering the cylinder the water is forced out the discharge outlet by the return stroke of the piston. This process requires valves to ensure that the flow is in the proper direction into and out of the cylinder. When the piston acts on the liquid in one end of the cylinder, it is a single-action pump. When the piston acts on both ends of the cylinder, it is a double-action pump, frequently called a duplex pump. Triplex and quadruplex designs are also possible. Duplex, triplex and quadruplex designs produce a flow that is smoother than that of the single-action model.

The *plunger pump* is a reciprocating pump with a plunger that does not come in contact with the cylinder walls but which enters and withdraws from the cylinder through packing glands. When the plunger is raised a vacuum is created below it, and water flows through a check valve to fill the void. When the plunger is lowered, the check valve closes, traps water in the pump, and forces it up through a valve in the plunger. Modern deep-well plunger pumps are a refinement of the old wood-cased hand pump used in country homes or farm wells.

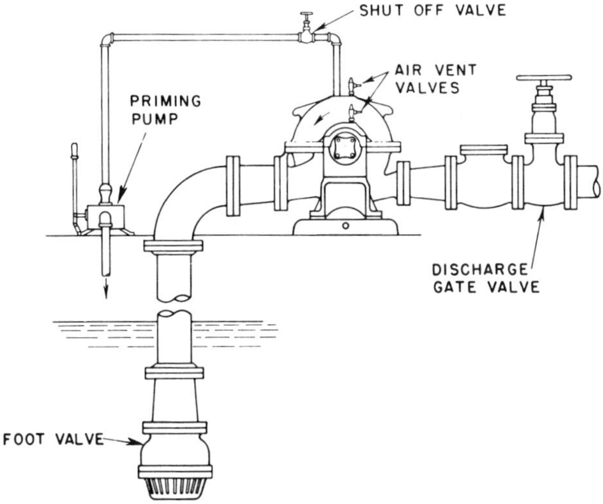

Fig. XI-11. Priming-Pump Installation

Diaphragm pumps contain a diaphragm as their operating part, which is usually made of a flexible material such as rubber. The diaphragm is connected to the edges of a cylinder. When the diaphragm moves in one direction a vacuum is created in the cylinder, drawing the water into the cylinder through a check valve. When the diaphragm moves in the other direction, the liquid is forced from the cylinder through a discharge valve.

Reciprocating displacement pumps are self-priming. Their flow rates may be adjusted by adjusting the strokes of the moving element of the pump.

Rotary-type displacement pumps (Fig. XI-13) consist essentially of close-fitting elements rotating in a pump case. These elements may be gears, lobes, screws, cams, rollers, or other devices. The movement of the elements is such that water is constantly being drawn into the pump through an inlet, then drawn through the pump and discharged. Rotary pumps act with neither suction nor discharge valves. The movement of the elements, rather than centrifugal force, causes the water to flow in one direction only. Rotary pumps will operate at any speed and produce a constant discharge at high pressures. However their construction usually precludes operation with water. They are better suited for more viscous fluids.

Other Pumps

Other types of pumps that would not be classified as centrifugal or positive displacement, are jet, air lifts, and hydraulic rams. *Jet pumps* are also called ejector-centrifugal pumps. This type of pump consists of a nozzle that discharges air, steam, or water into a constricted throat much like a venturi. This throat leads from a suction pipe. This arrangement permits the energy of a high-

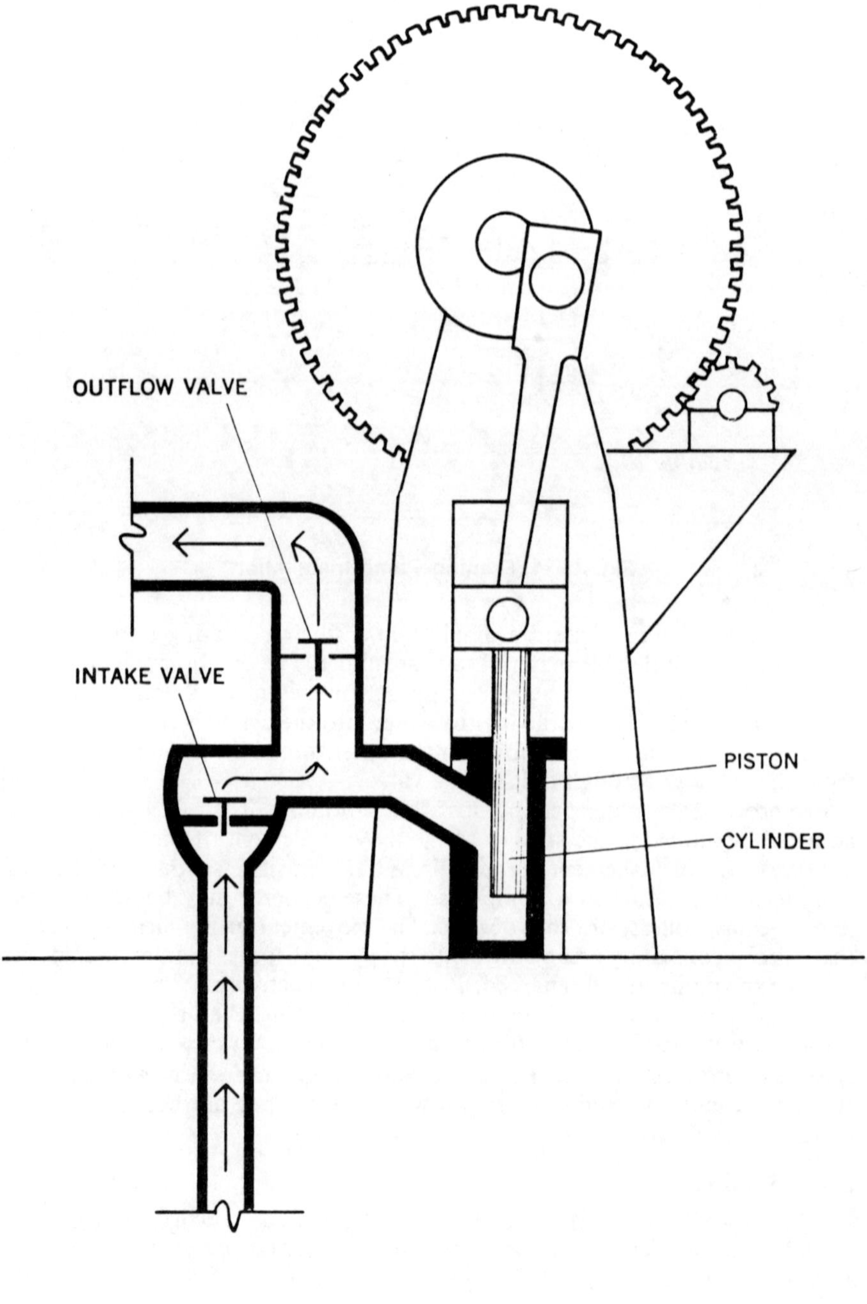

Fig. XI-12. Reciprocating Pump

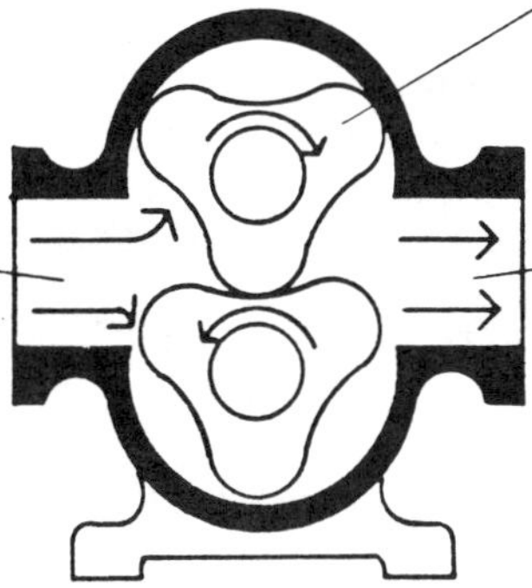

Fig. XI-13. Rotary Pump

pressure fluid (air, steam, or water) to be converted into that of a high-velocity fluid. With an ejector pump mounted on the top of the well, it is possible to lift water 400 ft. (Submersible centrifugal pumps can force water to yet higher elevations.) Because 35 - 50 per cent of the discharge goes back to the nozzle, this type of pump is not the most efficient, but it is frequently used in small wells for lifts greater than 25 ft and for net discharge of 50 gpm or more.

Air-lift pumps use a submerged air diffuser to lift water. When air as fine bubbles is introduced into water confined in a well casing, the water in the casing becomes lighter than the water outside. The heavier water on the outside forces the lighter water-air mixture up and out of the well casing. Although the air-lift pump is sometimes used to lift water from deep wells, it is not generally favored because it will not move the water horizontally through the distribution system. Also, because of the increased oxygen in the water, this method leads to a more corrosive water.

Hydraulic rams use the energy of falling water to raise a smaller quantity of water to heights greater than that of the falling water. This pump operates by taking advantage of the same forces that create water hammer in a system. Because of this, the hydraulic ram is often called an impulse pump. To operate correctly, this pump must have a constant supply of water available of the correct pressure. This is normally accomplished by having a reservoir feeding the ram. The quantity of water discharged by a hydraulic ram depends upon the working fall, the quantity of water flowing to the ram, and the total head or pressure against the ram. As a general rule, approximately one seventh of the volume of water flowing into the ram can be raised to about five times the height of the working fall.

Prime Movers

Electric, steam, gasoline, diesel, and natural-gas engines are used as sources of power for operation of reciprocating and centrifugal pumps. The electric motor has been the favored source of power to drive the pumps. This is largely due to the relatively low cost of electricity, reliability of electric motors, ease of automation, and generally quieter operation. Three types of electric motors are chiefly used: the squirrel cage, the wound-rotor (or slip-ring), and the synchronous.

The *squirrel cage motor* is the least expensive, simplest, and most reliable type of motor. It is used for constant-speed service.

The *wound-rotor motor* may be brought up to speed gradually, or the speed may be changed. Because of this, it is used to change the capacity of a pump by changing the pump's speed.

For large applications, requiring more than 150 hp, the *synchronous motor* is normally used. This motor is more efficient than the squirrel cage motor and uses less power.

Because of the popularity of electric motors, internal-combustion engines are seldom used as the main power unit. However, they are frequently used as back-up or reserve units or with portable pumps. As such they are much like the engine used in automobiles. Internal combustion engines are powered by diesel fuel, natural gas, or gasoline. Except for natural gas, these types of motors are not desirable for continuous use because of the high cost of maintenance upkeep and need for continuous fuel supply and upkeep. Natural-gas engines are often used because of lower fuel costs.

Steam and hydraulic turbines are still occasionally used as a power source. These units are reliable and efficient. However, because of the high cost of both the turbines and the accompanying boiler plant, they are seldom used today for any but older pumping plants.

Water Hammer and Cavitation

Both water hammer and cavitation have been discussed in previous chapters. These two physical phenomena can also be a problem in pump installations. However, proper operation will prevent major problems.

Water hammer may result from improper design, or the abrupt stopping of a pump, such as from a power failure. Because of the enormous pressures it creates, water hammer may rupture pipe or the pump case. Different methods may be used to prevent water hammer from either occurring or from causing damage if it should occur. Methods of control include a flywheel on the pump to slow starting and stopping, air chambers, surge tanks, surge suppressors, increment flow increase to the system, controlled-closing check valves, or controlling the motor speed.

Cavitation may result in destruction of pump parts. This problem often occurs when pumps are operated at speeds in excess of their designed rate. Noisy operation and vibration are usually evidence that cavitation is occurring or that the pump has been cavitated to some extent.

Operation and Maintenance

Satisfactory service from pumps can be obtained only through a proper preventative-maintenance program carried out by knowledgeable and attentive operators. These conditions must be accompanied by a sound installation.

Installation: Pumps are usually supplied with a bedplate under the pump and prime mover or under the pump only as shown in Fig. XI-14. The bedplate is de-

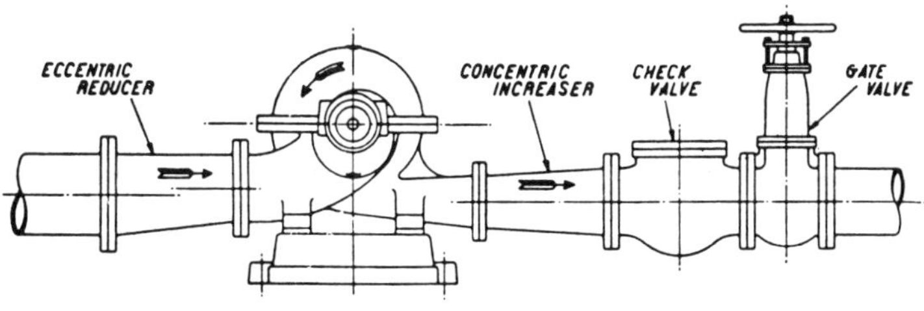

Fig. XI-14. Centrifugal-Pump Installation

signed to provide for the installation of anchor bolts and proper grouting to the pump foundation. These features are important since they seriously affect operating noise levels and alignment during the life of the equipment. Anchor bolts should be set into the concrete foundation and be of sufficient size as specified by the pump manufacturer. They should be grouted with a nonshrinking grouting compound so that no air voids remain between the bedplate and foundation. Ordinary concrete is a poor grouting material because of shrinkage.

Proper installation also requires that there be no strain on the pump case from suction and discharge piping. It may be desirable to have flexible pipe couplings in the suction and discharge lines to prevent pipe strain effects and to allow for slight changes in the pump setting after installation.

Suction piping for raw water should include a coarse-mesh strainer to prevent entrance of large foreign elements that could cause damage or improper operation. A pressure gage should be mounted on each side of the strainer to indicate by a drop in pressure that the screen needs cleaning and thus prevent pump damage.

Alignment: Alignment procedures for the pump and driver are critical and are required at installation even if units are supplied on a common bedplate, since the bedplate can bend or distort in shipment (see Fig. XI-15). Eccentric reducers should be used to connect the suction side of pump to piping. Since this reducer is level on top, it does not trap air. Alignment should be checked periodically to guard against premature bearing and coupling wear. Alignment should be checked when the pump is cold and again when the unit has run long enough for it to reach the proper operating temperature. On new installations the alignment can change considerably in a short period of time, and it should be checked daily until stable operation is established. Again, the pump or coupling manufacturer's specifications should be consulted for proper procedure and tolerances. In general, the coupling halves must be square and concentric with each other within certain limits depending on whether the coupling is of the solid or flexible type.

Priming: If the pump is operated under a suction head, care must be taken to vent any air from the top of the pump case. This can be done manually by the operator or by an automatic vent valve. If the pump operates with a suction lift it is important that a satisfactory priming system be used to prevent the pump from running dry. The priming system can consist simply of a foot valve on the

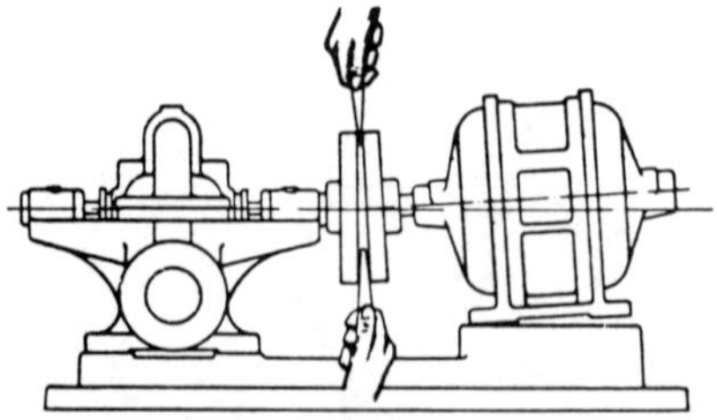

Measuring angular misalignment at top and bottom of coupling.

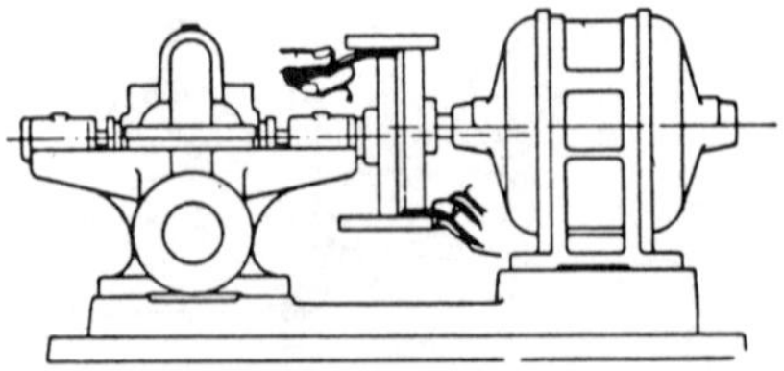

Measuring vertical misalignment at coupling halves.

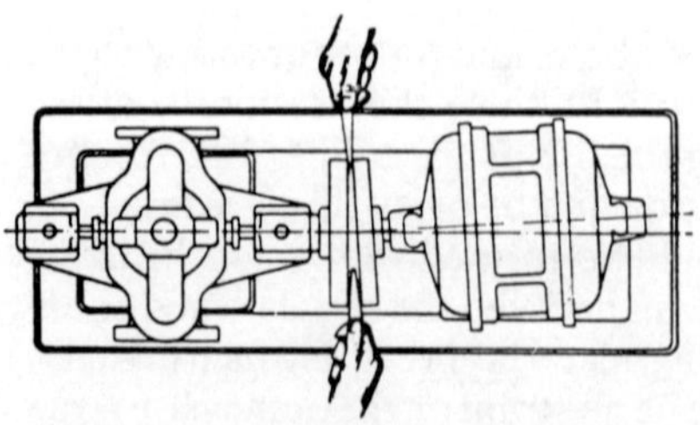

Measuring angular misalignment at sides of coupling.

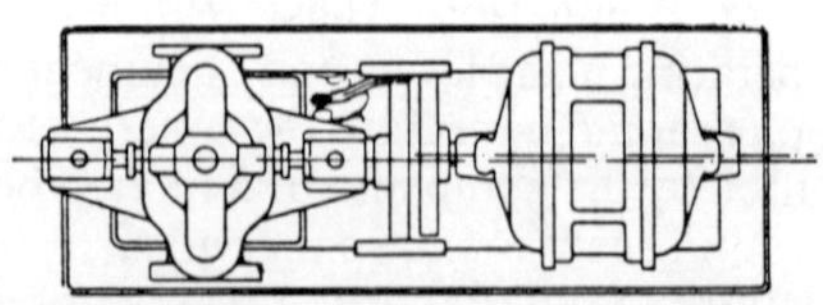

Measuring horizontal misalignment at coupling halves.

Fig. XI-15. Measuring for Pump Alignment

suction line to keep the pump full, or it can include a vacuum pump or ejector to draw any air from the system.

Packing: Before its initial start-up, the pump should rotate fairly easily by hand unless it is very large. If it does not turn over, the packing should be loosened until it does, provided there are no other problems. At start-up the packing should be allowed to leak freely until conditions stabilize and the operator has determined that there are no hot spots in the area of the packing glands or bearings. The packing gland can then be tightened very gradually and easily until it just drips. Packing glands should never be tightened to the point where there is no leakage, since this will cause premature packing wear and scored shaft sleeves. The leakage rate should be checked daily if possible.

When the packing wears or is compressed to the point where it is impossible to tighten the gland further, a new set of rings should be installed. The manufacturer's specifications should be consulted to determine the proper size, material, and number of rings required. It is not a good practice to add new rings on top of old ones in order to make up for wear and compression.

Bearings: Grease or oil lubrication should be adequate, but never overdone. Too much grease can cause a bearing to overheat, and too much oil can cause foaming. The best procedure is to follow the manufacturer's instructions and check the bearings periodically. Care should be taken to protect the bearing housing and lubrication cup from water spray or mist from the packing glands.

Chapter XII
Electricity, Instrumentation and Control

Introduction

This chapter is concerned with acquainting the operator with basic principles of electricity and automation. Much of the equipment the operator uses or is surrounded by is powered by electricity or controlled by automation. For effective use of this equipment, as well as safety for persons and equipment, the operator should understand the principles involved.

Terms

Electricity is like water in many ways. It is subject to flow, pressure, friction, and other water-like parameters.

Electricity, like water, is delivered under pressure. Electromotive force (EMF) is the force that gets the electric current flowing. Because the unit of measurement is the volt, EMF is commonly called voltage. It is usually represented by *E*, much as water pressure is normally represented by *P*. The voltage throughout a circuit can also be referred to as so many volts potential.

Electric current flow is measured in amperes, or amps (analagous to water flow being measured in gallons per minute). Current is represented by the symbol *I*.

Electric flow also has to contend with friction: The electrical terms are resistance and impedance, represented by *R* and *A*, respectively and measured as ohms (sometimes designated by the Greek letter omega Ω).

A watt is a measure of the power in an electrical circuit. The amount of wattage of a circuit is equal to the voltage *E* times the amperage *I*. Watts, represented by *W*, measure the total power in a circuit at any given moment, just as horsepower is a measure of the power produced by a motor at any given moment. One horsepower equals 0.746 kilowatts (0.746 thousand watts).

Energy is expressed in kilowatt-hours (number of thousand watt hours). For example 500 W for 2 hr or 1 W for 1000 hr equals 1 kWhr). The power company charges so many cents per kilowatt-hour.

A material, like copper, that readily permits the flow of electric current is called a conductor. Material such as rubber, which does not permit the flow of electricity, is called an insulator. Insulating material that is wrapped or cast around or otherwise surrounds a wire is called insulation. Insulation is normally used to prevent the flow of electricity from one conductor to another due to direct contact.

Kinds of Electricity

There are two principal forms of electricity, static electricity and dynamic electricity. *Static electricity* is an electric charge built up by friction, such as between two bodies or by rotating machinery, conveyor belts and the like. Discharging static electricity can be dangerous as it might cause a person to jump, causing an accident, or it can start fires or cause explosions. It can be prevented by providing proper electric connection (bonding) to ground or earth.

Dynamic electricity is electricity in motion. This generally refers to two types of motion—one in which the electric current with regard to a neutral reference maintains a consistent polarity (being positive *or* negative) and the other in which the electric current with regard to the reference reverses its polarity (being negative *and* positive) in a periodic manner.

Types of Current

Direct current (DC) is that current which maintains a constant polarity. In this case the voltage remains at a fixed potential across the path through which the current is flowing. Direct current can be developed by DC generators (rotating type) and can be obtained from, as well as stored in, batteries.

Alternating current (AC) is electric current that reverses its polarity in a periodic manner, rising from zero current to maximum strength, returning to zero and then proceeding through similar variations of strength in the opposite direction. In this case, the voltage across the circuit varies in potential force in a periodic manner similar to (though not necessarily in step with) the variations in current.

Alternating current is generated by AC generators (rotating type). One cycle of AC is shown in Fig. XII-1. The pattern shown is called a sine wave. If this wave-like action occurred once a second, it would be a 1-cps (cycle-per-second) or 1-Hz (Hertz) wave. The number of times this cycle occurs per second is called the frequency. In the US a frequency of 60 Hz is standard. AC current in the US reverses "direction" of flow 120 times per second (twice per cycle).

Three-phase current is nothing more than three separate AC waves on independent conductors arranged so that, at any instant, if the voltage is at a particular phase in one conductor it is one-third wavelength ahead in another and one-third wavelength lagging in the third conductor as shown in Fig. XII-2. Most large motors are three-phase, 60-Hz motors.

Ohm's Law

Ohm's law is a statement of the basic relation between voltage, amperage and resistance. Ohm's law states that the current *I*, measured in amperes (A), flowing in a given circuit is equal to the voltage *E*, measured in volts (V), divided by the resistance of the circuit *R*, measured in ohms (Ω). This relationship is usually represented by the mathematical equation:

$$\mathbf{E} = I \times R \text{ or } I = E / R$$

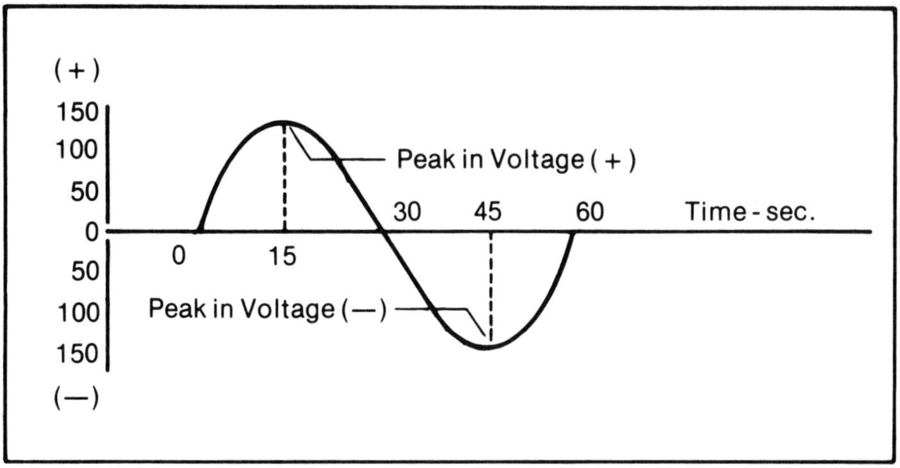

Fig. XII-1 Sine Wave Pattern of AC

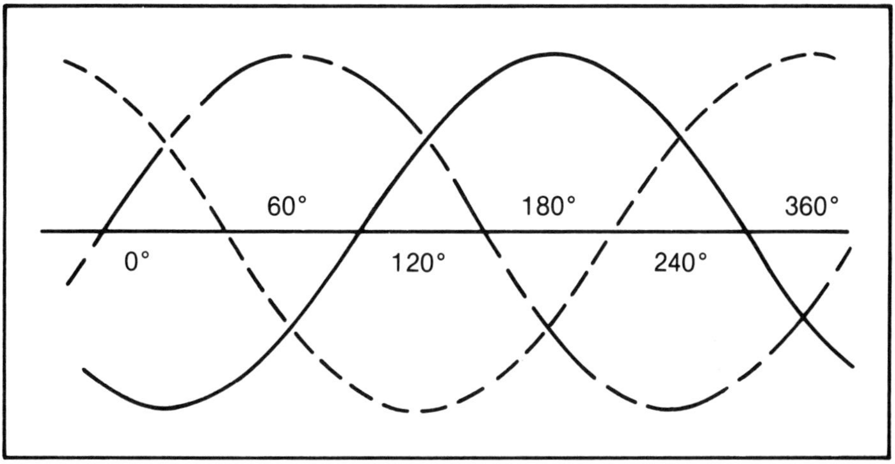

Fig. XII-2 Three-Phase Current

These expressions are sometimes shown by the "pie" diagram:

To use this figure, cover the unknown value and the remainder is the formula for finding it.

$$E = I \cdot R \qquad I = E / R \qquad R = E / I$$

A circuit impressed with 480 V and containing 100 Ω resistance would have 4.8 A current

$$I = 480 / 100.$$

The Ohm's-law equation confirms the water-electricity analogy expressed at the beginning of the chapter: For decreased pressure or increased friction, water flow will decrease. For decreased voltage E or increased resistance R, the current flow I will decrease.

Electric Circuits

Electric current will flow from any point of high voltage to a point of low voltage if there is an electric connection.

An electric circuit consists of the power source (a battery or generator); the load (the resistance in a circuit consisting of that which is heated, driven, or energized by the electricity); and the conductor, which must go from the power source to the load and return to the power source, completing the circuit. The circuit can be thought of as a pump (power source) forcing the flow (current) through a pipe (conductor) and back through the pump. Any apparatus (load) on this loop (circuit), which is driven by the water (current), adds resistance to the flow. The larger the pipe's (conductor's) size, the less the friction and the greater the possible flow. An open wire or switch in an electric circuit stops the flow as would a closed valve in a water line.

Types of Circuits

There are only three basic types of circuits. In the *series circuit* all resistances are in series. The current flow through one is the same as the flow through the other resistances, regardless of size. Water (current) flows through each apparatus (load) in turn. If the flow is shut off at any one point, it is off at all points. Introducing an additional load to the series circuit further reduces the flow. Each resistance to the flow contributes to the pressure loss (potential or voltage drop).

The *parallel circuit* has loads in two or more alternative routes similar to parallel piping. Each pipe (conductor) or apparatus (load) added in parallel effectively increases the size of the pipe (conductor), thereby reducing the resistance and increasing the flow. Even though the flow is cut off to one load, the flow continues through the parallel routes. Each parallel route is subject to the same pressure (voltage).

The third type of circuit is the *series-parallel circuit* and is a combination of the previously mentioned two. Parallel circuit rules apply in parallel portions; series circuit rules apply in the remainder. Parallel electric circuits are normally the type encountered by the distribution operator, so any equipment added to a circuit increases the electric current flow.

Often the return leg of an electric circuit is at ground potential or makes use of an earth-ground return. Shorting the circuit to ground or to the equipment chassis at any other point in the circuit would bypass normal resistance and might raise the current to a dangerous level, causing excessive heat in the delivery wires. A person simultaneously touching a live circuit and ground or equipment electrically connected to ground will become a conductor causing electricity to flow through his body.

Electrical Protection

The fuse is a common protective device in electric operation. A fuse is a short piece of wire or metal ribbon with a low melting point. During normal current demands, the fuse carries the current with no problem. However, during high amperage demands the fuse melts, breaking the circuit.

Circuit breakers are special switches designed to protect circuits. They may be operated either by heat or by electromagnetism. Upon overload, circuit breakers automatically switch to an off position, thereby breaking and protecting the circuit.

The thermal relay may be used to protect motors. This protective device senses the temperature generated by the current. When the current demand is sufficiently large to heat up the conductor, the relay trips, breaking the current flow.

The proper size and type of insulation are necessary to prevent shorting of wires. Also in some cases wire should be placed in conduit pipe for physical protection or guarding against damage from moisture. Conduit may be made of plastic (PVC) or metal. If metal conduit pipe is used, it may in some cases also be used as the equipment ground. Plastic conduit should have an extra wire inside to connect all exposed metal parts to ground. The equipment ground wire normally is bare or colored green. The neutral (usually white) wire to an electric service is also connected to ground.

Grounding

Ground is an expression representing an electric connection to earth or a large conductor that is at the earth's potential or neutral voltage. Motor frames and all electrical tools and equipment enclosures should be connected (called bonded) to ground. This is generally referred to simply as grounding, or equipment ground. Proper grounding will prevent a voltage charge from accumulating on the surface or within mechanical parts of any equipment.

The third prong on cords from electrical hand tools is the equipment ground and should never be cut off. When an adapter is used with a two-prong receptacle, the green wire on the adapter should be connected under the center screw on the receptacle cover plate. Many times equipment grounding, especially at home, is achieved by connecting onto a water pipe rather than a rod driven into the ground. This is not recommended because of the heavy use of nonconductor plastics and cement-asbestos water services and distribution lines. Care should be taken in breaking a service line or removing a meter because of the possibility of an electric charge being on the customer side. This is especially true where the meter is situated at the house or in a basement.

Generators

Lines of force run between north and south poles of a magnet. If a conductor moves across these lines of force, a current will be induced in the conductor, if there is a circuit for the current to flow. By rotating a loop between the magnetic poles, a current can be induced and a voltage generated. This is the basis of the electric generator. In practice, many loops and several poles may be used. How the current is picked off the windings of a rotating armature determines whether AC or DC electricity is produced.

Motors

Electric current through a conductor produces similar lines of force around the conductor as those between poles of a magnet. If this conducting loop is placed between two magnetic poles the loop will rotate when current is applied in order to align its north pole with the south pole of the magnet and vice versa. This movement can be continued by shifting the loop's magnetic poles electrically. In the simplest terms, this is the principle of electric-motor operation. The poles in a motor may be in the rotor (armature) or the stator (stationary winding).

Polyphase

The power of the single-phase AC line is fine for heating and lighting and powering small horsepower equipment. Because electric-power companies frowned on the use of large single-phase motors, they are not manufactured. Large single-phase electric power devices cause extensive power demand fluctuations. In two-phase operation the power sine waves are 180 deg apart. In three-phase power, the different phases are 120 deg apart. Most motors over 1½ hp used by water utilities are three phase.

Counter EMF

A rotating motor has wires cutting across lines of force, inducing an electromotive force opposite to that of the exciting voltage. This is called counter EMF and in effect adds resistance to the current flow. An energized motor that could not rotate would not have this counter EMF and would demand a very high current imput that would quickly burn out the motor.

Transformers

Because AC is basically on-off, the magnetic lines of force that surround a conductor expand and collapse, expand again with opposite polarity, and collapse again. If these expanding or collapsing lines of force cross a second conductor, the effect is as if the second conductor crosses stationary lines of force (see generators), and a current is induced in the second conductor, if it is part of a circuit, at the same frequency as the first's (see Fig. XII-3). Electrical devices making use of this concept are called transformers. The voltage in the induced (secondary)winding is in approximate proportion to the ratio of turns between the primary and secondary circuit. The product of the current and voltage of the two windings is about equal. If a primary winding with twice as many turns as the secondary is subject to 100 V and 6 A, the voltage generated in the secondary winding would be 50 V. Since the product of voltage-current is 600, the same in both windings, secondary current would be 12 A (50 x 12 = 600).

Transformers may be used to step up or step down voltage. Because high current flows require large conductors, usually the transmission of electricity is made at high voltage, reducing the current and the necessary conductor size.

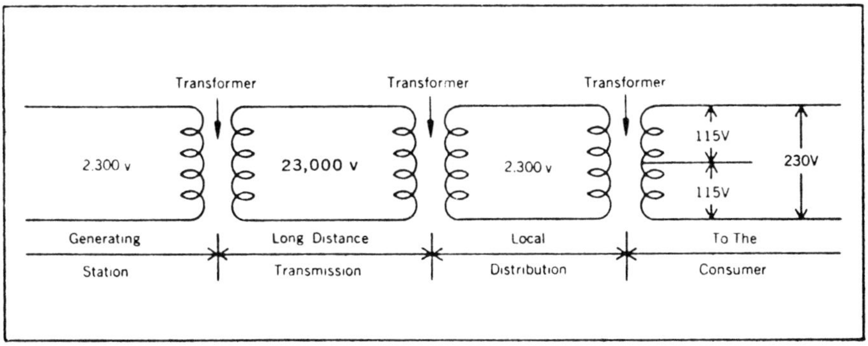

Fig. XII-3 Transformer Principle

Near the delivery point transformers are used to reduce the voltage to a safe-to-use level, increasing the current.

Transformers work only with AC: induction requires movement of the conductor or an expanding and collapsing electric field; DC with its stationary electric field would not induce current in the second conductor.

Relays

Relays make it possible to use a small amount of electricity to control a large electric flow. A small electrical coil drawing little current pulls in an armature which, when energized, closes contacts that turn current on or off to larger equipment. Before relays, motors had to be started by manually closing large switches. With relay starter-contactors, a small switch closes the relay contacts in the motor start circuit. Relays form the basis for automation as known today.

Relays and transformers are also used to isolate dependent circuits from each other. Without direct electric connection between circuits, one circuit can operate the other or signals can be sent from one to the other. Such isolation is common between phone company and customer circuitry. It is also used by power companies.

Capacitors and Coils

A *capacitor* (condenser) essentially consists of two or more parallel metal plates separated by insulators. Often this "sandwich" is spiralled in much the same fashion as a jellyroll in order to make the unit more compact. A voltage applied to a condenser will charge it to that voltage. The larger the capacitor, the larger the charge it accepts. Once charged, the capacitor will retain the charge until placed in a circuit that allows the charge to flow from the charged to the uncharged plates. Because with AC power there is an alternate charging and discharging of opposing charged plates, the capacitor, in effect, allows ready flow of AC but acts as a barrier to DC.

Coils (chokes) resist changes in current. This is due to the lines of force from one loop cutting across other loops in the same coil. An expanding electric field caused by an increasing electric current induces a counter EMF in the coil opposing that current, adding impedance (a form of resistance) to the current flow.

For this reason, coils resist a change in electric current, either up or down. Thus, coils add little resistance to flowing DC but add considerable impedance to continually changing AC.

Coils and capacitors, or a combination of the two, may be used to block or filter out unwanted AC or DC.

Rectifiers

Rectifiers block or partially block the flow of electrons (electric current) in one direction. This cuts off one half of the AC sine wave (as if a mask were placed over the lower half of a sine curve), converting alternating current to a pulsing form of direct current. Rectifiers can be combined in such a way as to provide full-wave rectification, which flops over the reverse portion of the sine wave so that all loops are in the same direction. Use of filters can smooth the pulsing DC. Rectifiers are also used to ensure that current is flowing in the proper direction.

Electronics

Electron tubes and transistors are electronic control devices that can be used as switches or devices that generate voltage or amperage in proportion to an input signal. It is possible to feed a small amplitude (strength) AC signal into an electronic circuit and have it act as a modulating valve, throttling the flow of electrons through the circuit. If the output is greater than the bias signal, the circuit is an amplifier.

Electron tubes and transistors are used for many other purposes than just amplification; some of the more common uses are as oscillators (frequency producing circuits), gating circuits (electronic switches), and rectifying circuits.

Solid state refers to transistors and some rectifiers, as well as some other electronic devices that have no air, vacuum or gas-filled gap between elements. Integrated circuits (IC) are complete solid-state circuits, including transistors, resistors, capacitors, etc., combined onto a small chip.

Instrumentation

Installation of adequate instrumentation and controls is one of the better ways of assuring reliability of water-system operation.

The term *instrumentation* is usually applied to the measurement of process variables such as flow, pressure, reservoir level, and chlorine residual, whereas the term *control* implies regulation of equipment to produce the desired effect.

Most of the instrumentation and control in water system operations is used for the measurement and control of flow, water level, and pressure. In addition, more and more water-quality parameters are being monitored, such as chlorine residual, turbidity, temperature, and pH.

Use of Instrumentation

Some typical instrumentation and control applications are

Pumps
1. Flow metering at each station
2. Alarms (activated in case of malfunction) for power failure, pump failure, and other malfunctions
3. Automatic operation

Chlorination	1. Automatic chlorine proportioning, residual measurement, and recording
	2. Automatic operation of fans and lights
	3. Failure alarms
Reservoirs	1. Elevation or amount in storage
	2. Water flow in and out
Canals	1. Flow
	2. High- and low-water alarms
Distribution System	1. Pressure at critical locations
	2. Pressure regulation

Control and Automation

A typical control system consists of (1) the process to be controlled; (2) a device for measuring the variable; (3) a transmitter; (4) a controller, which compares the measurement with the set point (desired setting) and transmits a control signal to (5) a control element, (such as a valve) that effects the necessary change.

Control can be accomplished either manually, semiautomatically or automatically. In manual control, the operator performs individual control functions either by turning valves, closing switches or pressing buttons. In semiautomatic control, the operator manually adjusts the controller, which then automatically performs the programmed control sequence. In automatic control, the controller is actuated automatically by a signal from the process.

The simplest automatic control is the on-off switch that has only two positions, such as valve open or closed. The home thermostat is a familiar example of on-off control. On-off is also used in pumping stations to start or stop pumps according to the reservoir level or pressure. For control system examples see Fig. XII-4.

Control Loops

Two types of control loops, the open and the closed loop, are used in automatic control. In the *open-loop control,* the controller varies the supply without ascertaining the effect. For example, most proportional-feed chlorinators use open-loop control: they proportion chlorine to water flow, leaving out the measurement of chlorine content. In *closed-loop control,* feedback measurement of the effect of the process is added to the circuit so that the controller can compare the treatment operation with the set point and make any necessary adjustments. Control of chlorination by means of a residual chlorine analyzer is an example of a closed-loop system. Here the information from the resulting chlorine residual is fed back to the controller, which then increases or reduces the chlorine dose to achieve the desired residual. This will be further explained later.

Pumping and rapid sand filtration are the two processes for which automatic operation is the norm. Pumping may be controlled by the water level in a reservoir or by water pressure. The filter flow rate may be held at a constant level regardless of build up of material in the filter bed, or the flow may be regulated by the water level of the clear-well or storage basin. Filter backwash is usually semiautomatic: the operator initiates the backwash sequence. However

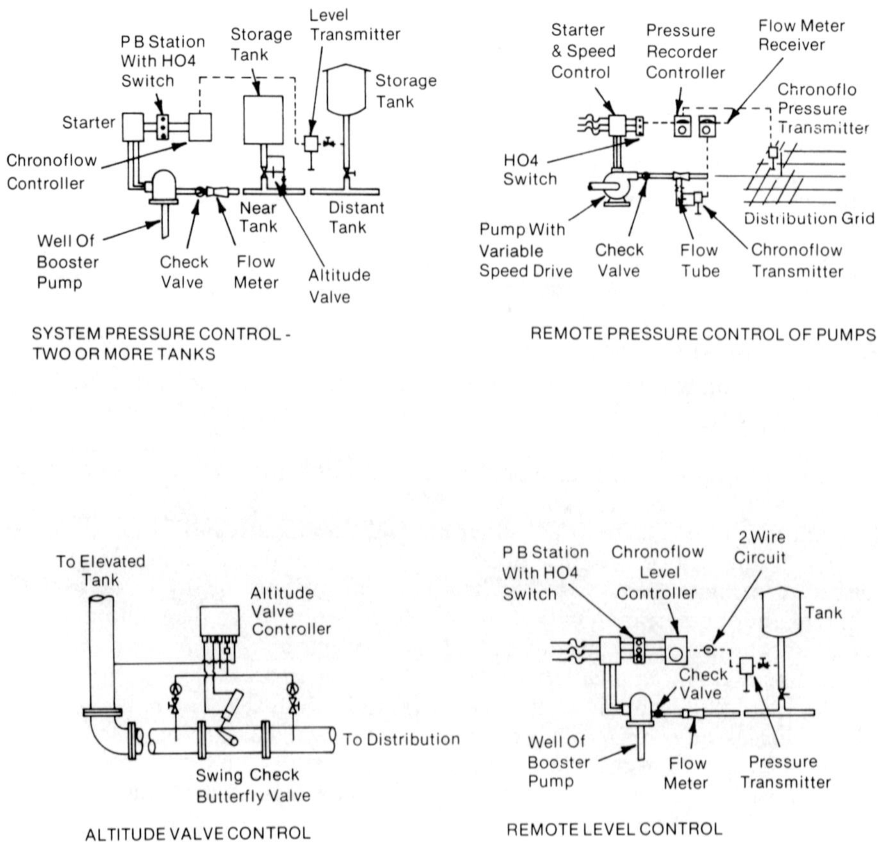

Fig. XII-4 Examples of Control Systems

automatic systems actuated by head loss, turbidity, or other continuous measurement are in operation.

Pumping Stations

Automatic control of pumping stations is a universal practice. Water level in reservoirs supplied by most old stations is controlled by constant speed, on-off devices that may be either float- pressure- or electrode-actuated. Pumping rate is increased stepwise by switching on additional pumps or by changing pump speed if variable speed pumping units are used.

In variable-speed-pumping control, the pumps are driven by a variable-speed drive, such as an engine or adjustable clutch, which is varied by a controller.

Other control elements used at pumping stations include pump-sequence programmers; mechanisms for automatic transfer to emergency power sources (fossil-fuel-engine starters and transfer switches); low- and high-water-level alarms; mechanical-trouble signal lights; timers to provide lag time between power failure and transfer to emergency power; and hour-run meters to determine preventive-maintenance periods.

Control of Chlorination

The application of chlorine for disinfection can be controlled by a variety of methods, which depend partly on the type of feeder used. Most control parameters involve chlorine dosage adjustment to achieve a desired chlorine residual. However, some include safety features to protect the operator from exposure to chlorine and to prevent interruption of the disinfection process.

Mechanical diaphragm feeders: The mechanical diaphragm feeder dominates the chlorinator field. These feeders are characterized by spring-loaded, mechanical, diaphragm-operated, pressure-reducing valves; rotameters for indicating rate of feed; diaphragm-operated, vacuum regulators downstream from the rotameter; safety shutoff valves; and vacuum gages for indicating proper vacuum in the system.

The chlorine feed rate is varied either by adjusting a needle valve or a proportional valve. In order to change the chlorine feed rate automatically, the valve plug is positioned by means of an actuator such as a small electric motor ordiaphragm air motor.

Control of chlorine dosage: Chlorine dosage may be controlled by any of the following methods:

1. Manual
2. Remote-manual
3. Intermittent start-stop
4. Program rate
5. Open loop, flow proportional
6. Closed loop, chlorine-residual response
7. Compound-loop control

In manual control, starting, stopping and feed-rate adjustments are accomplished manually at the chlorinator. In remote-manual control, these operations are performed manually at the control panel by means of switches, push buttons and indicators. In start-stop control, chlorine feed rate is adjusted manually, but starting and stopping are accomplished automatically. (An application would be operating a feeder in conjunction with a pump operation.)

The start-and-stop operation is usually accomplished by starting and interrupting the injector water-supply pump.

Flow proportional dosage control: In flow-proportional control, the chlorine feeder is varied automatically in proportion to flow rate (see Fig. XII-5). This is an open-loop system without feedback from the chlorinated water. The control mechanism is similar to that used in other systems; i.e., the chlorinator is controlled by a signal from a flow meter or differential-producing orifice plate in the water line. Chlorine feed rate is varied in proportion to flow. However, the prescribed dosage is preset manually.

Flow-proportional chlorine dosage control is widely used in water treatment. The associate equipment is reliable, efficient, and inexpensive. The system provides for economic use of chlorine in applications where the chlorine demand is fairly constant.

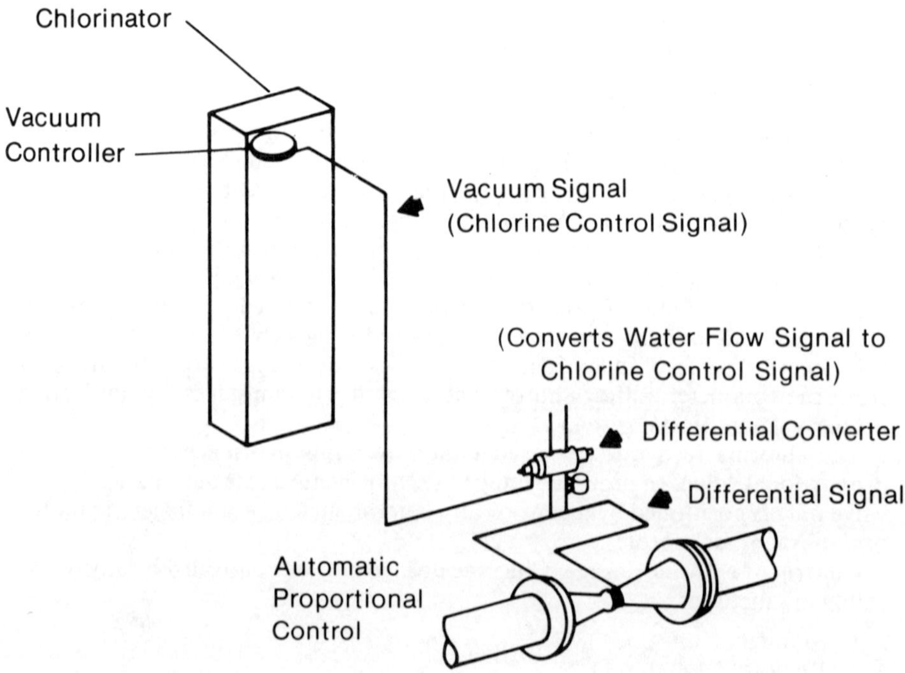

Fig. XII-5 Flow Proportional Dosage Control

Closed-loop control: In closed-loop control, chlorine dosage is varied in response to chlorine demand, as determined by a feedback chlorine-residual measurement. (See Fig. XII-6.)

This type of control depends on the use of an automatic chlorine-residual analyzer: the chlorine-residual measurement signals are fed back to the chlorinator, and the chlorine feed rate is adjusted automatically to maintain a predetermined chlorine residual.

Compound-loop control: In many cases of closed-loop control (Fig. XII-7), the chlorine-residual control loop is combined with a flow-responsive control, giving rise to a compound loop. In compound-loop chlorine-dosage control, the flow transmitter and a controller adjust the chlorine feed rate in a fixed ratio to the flow, as described in a previous section. The chlorine residual analyzer provides compensation for changes in chlorine demand, usually by means of a separate controller. Thus, constant chlorine residual is maintained under conditions of both variable flow and chlorine demand. A typical compound-loop chlorinator control is shown in Fig. XII-7.

Alarms: Alarms that give a warning in the event of chlorinator malfunction or deviation from set operating performance are extremely important in water disinfection. Special chlorinator alarm units, usually electrical, that provide both audible and visual signals are available. The audible buzzer, horn, or bell can be manually silenced, but the light will remain on until the malfunction is corrected. A red light usually indicates trouble, and a green light or amber light

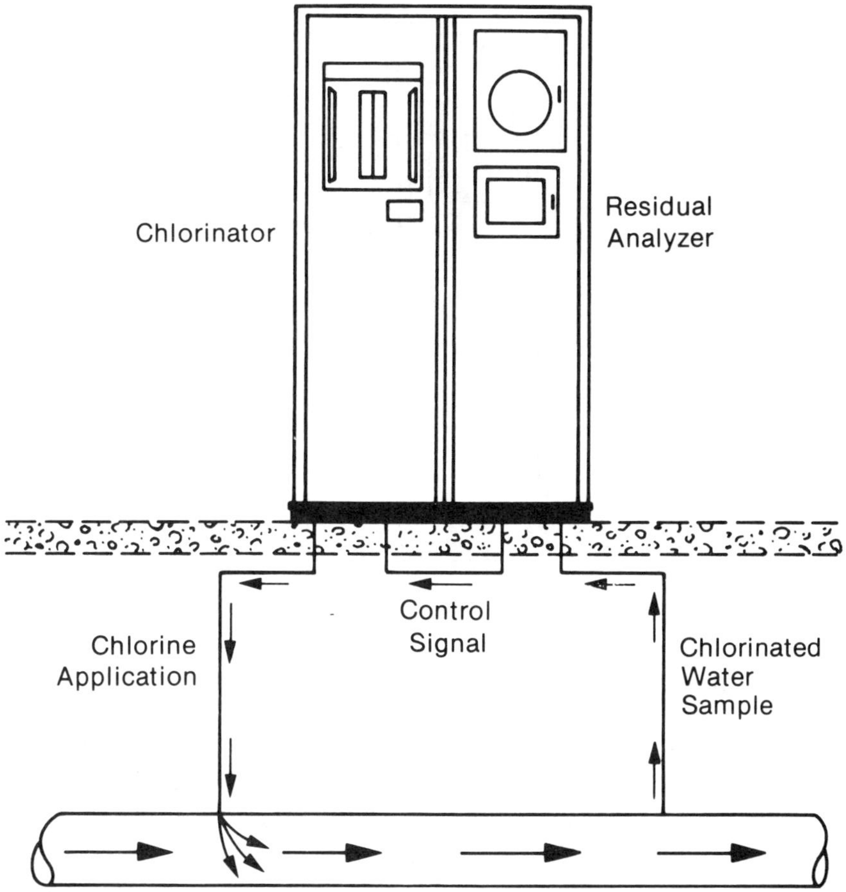

Fig. XII-6 Closed-Loop Control

indicates proper operation of both the chlorinator and the alarm system. Any trouble spot should be identified by letter markings near the alarm light or by backlighted script that illuminates in case of trouble.

Alarm systems can be energized at a number of signal points to indicate trouble, including the following:

1. High- and low-temperature thermostats on liquid-chlorine evaporators
2. Low-pressure switches on water supply
3. High- and low-pressure switches on chlorine supply
4. Scale dial switch indicating empty chlorine cylinder
5. Chlorinator vacuum switches for high or low vacuum, indicating injector clogging, failure of injector water supply, loss of chlorine gas supply, or other difficulty.
6. Alarm switch on chlorine flow recorder, indicating exhausted chlorine supply or failure of injector water supply

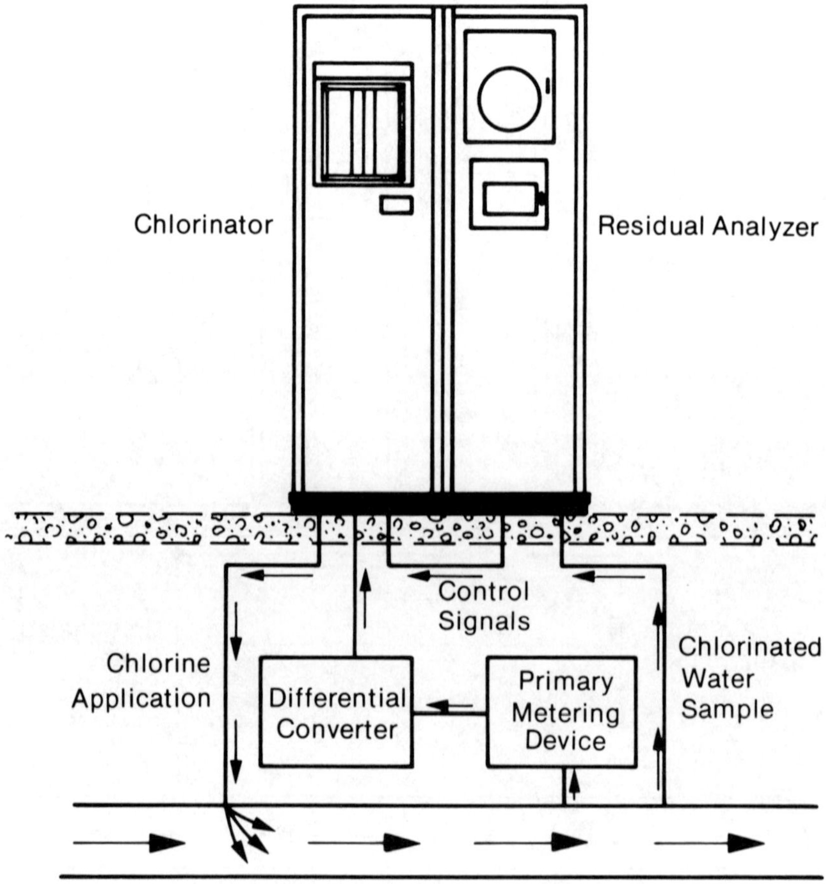

Fig. XII-7 Compound-Loop Control

7. Limit switches indicating chlorine residuals outside preset limits
8. Low-temperature in chlorinator building
9. Water leak
10. Chlorine leak

A chlorine-pressure-regulating valve or a shutoff valve can be connected into the alarm system so that the chlorine supply is shut off automatically if certain alarm circuts are energized.

Chlorinator alarms are in common use at water-treatment plants. Alarms can be triggered at either high or low vacuum in the chlorinator. High vacuum indicates interruption of chlorine supply, and low vacuum indicates injector-water failure. The alarm signal can be transmitted by wire or radio to the operations office and possibly via radio to all of the district's vehicles. In addition, the alarm can automatically start an emergency hypochlorinator. Both the radio transmitter and the hypochlorinator can be battery powered, which allows them to operate unaffected by power failures.

Auxiliary control features. Additional features that have been recommended in chlorination control include:

1. Self-starting standby power sources
2. Automatic devices for switching over to a full chlorine cylinder
3. Equipping of scales for weighing chlorine with loss-of-weight recorders
4. Chlorine gas pressure gages to indicate lowered tank content or freezing of line
5. Automatic operation of fans and lights in chlorinator room

Remote Monitoring and Control

Remote monitoring and control of widely dispersed systems from one or more central points is usually called telemetering. Telemeters are long-distance transmitters and include both the old electric and the new electronic types.

Most telemetering systems in water control today are essentially monitoring systems. However, automatic or semiautomatic control elements are available for converting a monitoring system into a remotely controlled system. All quantities that normally can be controlled directly can be placed under remote control; e.g., motors, circuit breakers, valves, compressors, etc. All significant quantities can be monitored; e.g., pressure, valve position, reservoir-level readings, motor on or off, flow readings. The proper transducer or converter will change the quantity into an electric signal. The telemeter signal can be as simple as an on-and-off DC signal. In control, the DC current might be of one polarity for opening a valve and the opposite polarity for closing.

Other signals used in monitoring include a single electric frequency assigned to one condition, such as low water level.

All primary measuring devices produce an analog movement; that is, a signal proportional to the measured value of flow, valve position, pressure, etc. Analog signals are proportional signals. A condition is indicated by varying the width of pulses (pulse duration modulation, PDM), the number of pulses, voltage, current, or frequency. PDM signals generally will last a specific number of seconds in a 15-s cycle. Voltage signals may be 0 - 0.1 V, 0 - 1 V, etc. Current signals are usually in a range of five to one, such as 1 - 5 mA (milliamps), 4 - 20 mA, etc. A PDM signal can be generated by use of a cam as shown in Fig. XII-8.

An analog signal may be converted to a digital code signal with an analog-to-digital converter. A digital code is a form of counting. The digital signal is fast, accurate, and of a form that can be used directly by computers. Monitoring signals are displayed on the control panel by indicators, recorders, or signal lights.

Telemetering signals are transmitted via leased telephone lines, private cables, microwave or radio systems, or carrier systems on utility-owned powerlines. The use of telephone lines necessitates the conversion of measured signals to audio-tone frequencies.

Data processing equipment, including scanners and digital data loggers, are used at large, complex systems. Scanners are used to monitor and record large numbers of measurements. Data loggers are used to record, store, and analyze system operating data. (Fig. XII-9) The data can be fed to the data logger either

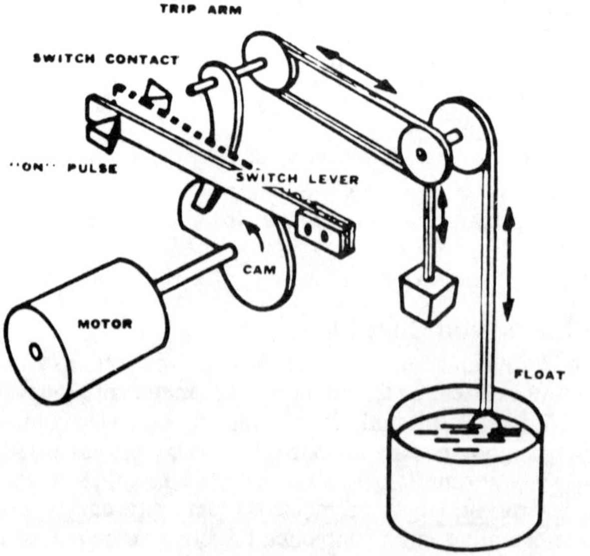

Fig. XII-8. Transmitter

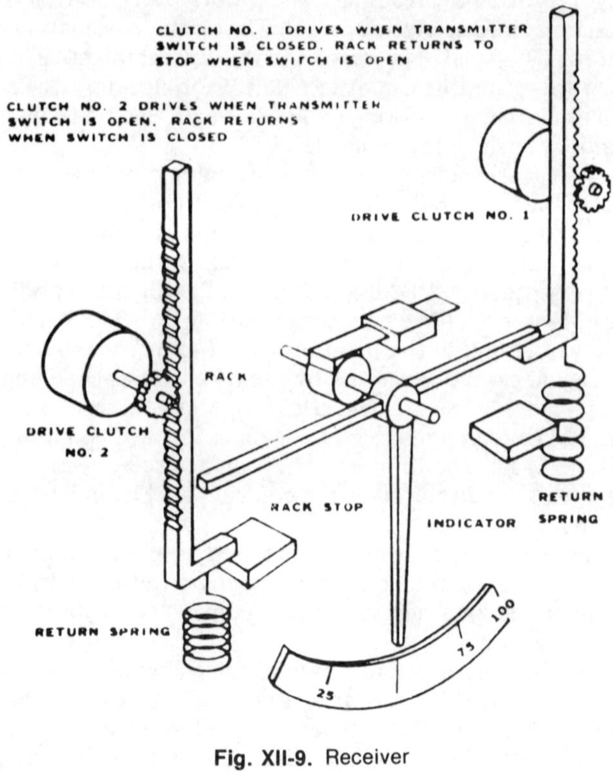

Fig. XII-9. Receiver

directly from a scanning system or manually by means of a typewriter. The loggers can be programmed to print out the data at regular intervals, to do a limited amount of statistical analysis, and to prepare monthly or annual operational reports.

Control Panels

Control panels are used for compact assembly of recorders, indicators, switches, and alarms at a convenient central location. Graphic panels show system elements, control points, and operating conditions on a map or hydraulic diagram with miniature multicolored lights. These panels may also contain indicators, recorders, and controllers.

Panel alarm systems consist of alarm monitors and special alarm annunciator panels. The annunciators indicate the trouble at any part of the system by signal lights, visual markings, and/or audible alarms. The alarm is usually stopped manually, whereas the light stays on until the trouble is corrected. Alarm systems are often provided with automatic shutdown of equipment.

Supervisory Control

Supervisory control is the selective control of remotely located devices and confirmation of their performance. Supervisory control equipment includes remote monitoring (telemetering) and control elements, alarm systems, data loggers, computers, and control panels. The function of supervisory control is to monitor and control dispersed systems effectively from one or more locations. Supervisory controls used in water-utility operation may range from simple alarm annunciator panels with start-and-stop control to sophisticated computer systems.

Control signals available are continuous, momentary, and proportional. Any of these signals can be transmitted by an operator or computer or by feedback from monitoring devices.

Continuous signals are used in electronic transmissions and have already been discussed. A momentary signal is generated by the operator pressing a button. This keys a frequency generator or a coded series of pulses that is sent to the distant receiving station. The momentary signal is "remembered" at the receiving station by a mechanically or electrically latching relay, magnetic core, or other memory device that performs the required operation such as starting a pump. Proportional control is accomplished by maintaining the signal as long as the action is wanted, as for positioning a valve be means of a series of push buttons such as on, off, high, low, etc. The results are fed back to an indicator for comparison by the operator.

Data Processing

Data processing involves the monitoring, recording, computing, logging, and evaluating of variables in order to obtain improved operation, protection of the equipment, and saving in time. The computations range from simple scaling to operating guides. In computerized systems, the computer may be applied to closed-loop control of the process.

In continuous monitoring, as discussed previously, all points are observed continuously. Each variable being monitored must have its own instrumentation and detection equipment. Scanning systems are programmed to observe or sample sequentially a large number of points, whereas the recording and indicating equipment is time-shared, thereby reducing costs.

The objectives of data processing are: (1) to provide alarms in case of process or equipment malfunctioning, (2) to supply accurate records for accounting, (3) to provide records for process analysis, (4) to compute operating guides, (5) to accomplish automatic control, and (6) to provide for automatic adjustment of mathematical models.

Digital Data Loggers

Digital data loggers contain all the functions of the monitoring and recording system and can also provide some analog computation and digital records such as typewriter printout.

The logger is programmed to print out all variables at specified intervals (e.g., hourly,). Between logging cyles, all variables are scanned for off-limit values and printed out in red when detected. Another printout is made when the variable returns to normal.

Digital Computer Control

The use of a digital computer affords completely automatic control. Specifically designed digital computers are available and are being used in automated water systems. The advantages of computer control have been reported as follows: (1) decreased operating cost, (2) increased system efficiency, (3) increased safety of operation, (4) rapid logging and summarizing of operating data, (5) ability to perform special tasks such as preventive-maintenance scheduling and plant-inventory control, and (6) ease of expansion to provide for additional plant capacity.

Maintenance

The fact is that operators see more of the equipment than technicians. The operator must not only know how to use the instruments and how to get the greatest amount of information and control from them but also how to recognize malfunctions before they are serious. He is usually expected to take simple everyday care.

Some of the things an operator should observe and report are

1. Any unusual heat, odor, or noise
2. Pen hunting (constant up and down movement of chart pen)
3. Inconsistent indication (different readings when approaching a particular reading from a lesser value than when approaching from a higher value)
4. Disagreements among readouts (dials, pointers, charts, and totalizers)
5. Instrument jerky in movement
6. Disagreement of instrument with other information
7. Pen inking too little or too much
8. Chart pen not always on correct time
9. Any improper dial readings
10. Fluttering indicator lamps or relays, which indicate a poor connection or weak signal

Sometimes there are no technicians available, and the operator is expected to do more maintenance. Regardless, an operator can do much to aid in the maintenance program to help keep the equipment functioning properly and accurately. Some operation and maintenance procedures to be observed are

1. During chart changing pen should not be moved.
2. Clogged ink pens usually can be cleared by squeezing plastic ink reservoirs; if this fails a small spring can be used to apply pressure to the pen or tubing.
3. Instrument face should not be wiped with a cloth unless it is treated because it may magnetize the indicator.
4. If oiling, proper instrument oil should be used without overfilling.
5. Program should exist for periodic operation and testing of controls.
6. Air lines should be bled of moisture.
7. Water lines should be bled of air.
8. Pressure lines to instruments should be flushed with a large volume of water.
9. Screens should be cleaned periodically.
10. Air filters should be replaced according to a schedule.
11. Indicator lamps and audio alarms should be tested.
12. Water flow or air flow to instrument should be correct.
13. Temperature compensation setting should be adjusted.
14. Reagent reservoirs should be filled.

Precision instruments and controls must be used with care. They serve properly only when they are working correctly. They are great tools for operators and should not be left in a malfunctioning condition since this will cause loss of confidence in their use.

Chapter XIII
Safety

Accidents do not just happen, they are caused. This chapter is intended to provide water-utility employees with a background in the areas that are of special concern for their type of work. This is by no means a complete safety discussion. For those who wish to cover this area more thoroughly, *Safety Practices For Water Utilities,* Manual M3, is recommended.

Responsibility

Everyone is responsible for the maintenance of safe working conditions. This includes the company, the supervisor, and the individual. The company, through the management, is responsible for maintaining both safe working conditions and supporting a policy that encourages safe performance of work duties. Supervisors at all levels are responsible for the direct control of work conditions. This means that each crew chief has a responsibility to see that all work is done in compliance with safety practices and regulations.

The employee has a special position with regard to safety. Safety practices and equipment must be used, or the safety program will not be successful. The employee must help to see that unsafe acts and conditions do not occur.

Regulatory Agencies

Through the Williams - Steiger Occupational Safety and Health Act of 1970 (OSHA) the federal government has established minimum health and safety standards. Every employer must furnish employees with a place of employment that is free from recognized hazards that are likely to cause death or serious physical harm. The act provides for six months in prison and/or $10 000 penalty on conviction.

Under the provisions of the act, each state may take over the enforcement if it establishes standards and enforcements that at least meet the minimum standards of the act.

Causes of Accidents

As previously stated, accidents do not just happen; they are caused. Often, however, the cause of an accident or a potential accident may not be readily apparent. However, the cause can normally be discovered using common sense, personal observation, inspections, and accident studies.

An accident can almost always be traced to either an unsafe act and/or to an unsafe condition. All workers should learn to recognize both unsafe conditions and unsafe acts.

Personal Protection Equipment

Most utilities now provide a broad range of personal-protection equipment, which is normally quite successful in helping to prevent serious injuries. Hard hats, safety goggles and steel-toed safety shoes are perhaps the most widely used safety equipment.

Face protection and eye protection are possible through the use of *goggles and face shields.* These are required where there is a chance that injury can be prevented by their use.

Steel-toed safety shoes and boots are helpful in preventing damage to feet, which can be caused by dropped objects. Where more protection is needed from very heavy impacts, such as when operating tamping equipment, then thick-flanged and *corrugated-sheet-metal foot guards* are recommended.

Gloves are necessary for protection from rough, sharp or hot materials. Special long-length gloves are available to provide wrist and forearm protection. Rubber gloves should be worn when handling acids, alkalines, oils, solvents, and other chemicals. However, gloves should not be used around revolving machinery.

Hard hats have been very successful in reducing serious injuries or deaths due to head injuries. Hard hats are required whenever a person is working in a trench or wherever a person can be injured by falling objects. Metal hard hats should not be used where there is an electric hazard.

Whenever employees are subjected to certain noise exposures, (see chart) they must be provided protection against the effects of the noise exposure.

Table XIII-1
Permissible Noise Exposures

Duration *hr/day*	Sound Level *—dB*
8	90
6	92 - air compressor
4	95 - air drill
3	97
2	100 - machine shop
1½	102
1	105 - circular saw
½	110
¼ or less	115 - diesel engine

Respiratory equipment is frequently needed in the water-utility field. Respiration devices provide protection from dusts, mists, fumes, vapors, and gases. Care should be exercised that the proper type of respirator is used for each condition.

Filter respirators fit the lower part of the face. They mechanically remove dust, fumes, and mists. Each filter is designed for a particular application and should be changed when breathing becomes difficult.

Canister and cartridge respirators, including gas masks, use chemical action to neutralize or filter out the contaminant. This type of respirator is seldom approved for concentrations greater than 2 per cent for most gases and should not be used where there is inadequate oxygen in the air. These respirators should be occasionally checked to ensure that they are in good operating condition.

Supplied-air respirators consist of either a helmet or a face piece covering the nose and mouth. Air is then supplied through a hose from an approved source. These devices are frequently used for work in tanks.

Air or oxygen breathing apparatuses are used in oxygen-deficient areas or where the concentration of dust or gases is extreme. These units either have a compressed-air cylinder or chemicals in a canister, which generate oxygen from a reaction with the moisture in exhaled breath. All persons who might need to use these apparatuses should be instructed in their use (see Chapter XIV).

Oxygen and gas detectors are available for testing the air in manholes and tanks to ensure that it is safe to enter. These devices sample the atmosphere continuously and provide warning if the air is unsuitable.

Safety belts are protective devices designed to provide support and to limit any fall. They are frequently used by persons working in high places on insecure platforms, and in other hazardous situations.

First-Aid Facilities and Training

All injuries should receive first-aid attention and, if serious, prompt attention by a physician. This is only possible if everyone receives first-aid training (see appendix). Excellent first-aid training is normally available through a local Red Cross organization. Also, the local Heart Association will normally provide training for cardiopulmonary resuscitation. This training is especially important for persons who work around electrical equipment or chlorine.

All individuals should know the location of first-aid equipment. The location should be well marked. All first-aid kits should be properly supplied and maintained.

Manual Handling of Material

Everyone should know the proper way to lift objects by hand. The correct procedure is: (1) get a good footing with feet about shoulder width apart; (2) bend at the knees to grasp the weight, keeping the back straight; (3) get a firm hold; and (4) keeping the back as straight and upright as possible, lift slowly by straightening the legs.

Whenever it is desirable to change directions, the whole body should be turned, including the feet, rather than just twisting the back. No one should ever lift a load that is too heavy or too large to lift comfortably.

When transporting objects, whether by a wheelbarrow, hand truck, or cart, care should be exercised. Equipment should be properly maintained and not overloaded. Clearances should be carefully judged, and horseplay should be prohibited.

Structural Hazards

Scaffolds, ladders, stairways, walkway aisles and handrails can be structural hazards. All scaffolds should be maintained in a safe condition. Unstable objects such as barrels, blocks, boxes, or loose brick should not be used to support scaffolds or planks. Any scaffold damaged or weakened from any cause should be immediately repaired and not used until the repairs have been completed. Scaffolds should not be loaded in excess of the working load for which they are designed.

Tools, materials, and debris should not be allowed to accumulate in quantities to cause a hazard.

An individual ascending or descending a ladder should face it. Only one person should be on a ladder at one time. The top of a stepladder should never be used as a platform on which to stand. Straight ladders should be placed so that the distance of the base of the ladder from the wall is one foot for every four feet of the ladder. Everyone should make a habit of carefully inspecting any ladder before using it.

Stairways should always be kept clear of tools, supplies, and other materials. Also, grease, water, and other slippery substances should be kept off stairs. Running on stairs and not using handrails should be discouraged. All stairs should be properly lighted.

Aisles and walkways should be kept clear of obstructions. Handrails should be of adequate srength and firmly mounted. All stairways wider than 80 in. should have a middle handrail as well as a railing along the edge.

Hand Tools

Basic rules for hand-tool use are that personnel

1. Always select the appropriate tool for the job, never a makeshift.
2. Check the condition of a tool at frequent intervals, and repair or replace it if defective.
3. Avoid using tools on moving machinery; stop the machine before making adjustments.
4. Check clearance at the work place to make sure there is sufficient space, to recover it if the tool should slip.
5. Have good support underfoot so that there is no hazard of slipping, stumbling, or falling.
6. Protect hands by wearing gloves in confined places, except when using hammers. (Rings should not be worn.)
7. Carry sharp or pointed tools in covers, or be sure they are pointed away from the body in case of a fall.
8. Wear eye protection when using impact tools on hard, brittle material.
9. After using tools, wipe them clean and put them away, thus keeping the work place orderly.
10. Do not lay tools on top of stepladders or on other places from which they could fall on someone below.
11. Learn and use the right way to work all hand tools.
12. Use tools made of nonsparking material where fire is a hazard.

The right tool should always be used for the job being done. Files and screwdrivers should not be used as pry bars. Wrenches should not be used as hammers. A proper size of wrench should be used rather than a pair of pliers.

All tools should be kept in good condition. Chisels with mushroomed heads, dull saws, hammers or shovels with cracked handles, and wrenches with spread jaws should not be used.

All tools should be used properly. Wood chisels should never be pushed toward any part of the body. Screwdrivers should not be used on objects held in the hand.

All tools should be kept in a safe place. They should not be left on overhead surfaces or ladders. Tools, especially ones with sharp edges, should not be carried in trouser pockets. Workbenches should be neat and orderly.

Portable Power Tools

Care must be taken when using *electrically powered tools* to avoid dangerous shocks. Insulating platforms, rubber mats and rubber gloves should be used when additional care must be taken to avoid shocks. Electrical tools should always be grounded when in use. The cord should never be used as a line to raise or lower the tool. Eye protection should be worn if there is any danger of flying particles. Cords should always be carefully maintained, and, when in use, they should not create a hazard of tripping or stumbling. Electrical tools, if properly maintained and operated, need not be the cause of an accident.

Pneumatic or air tools are similar in operation to electrical tools. Hoses should be protected and properly maintained. Safety-type hose connections should always be used. Pneumatic tools should neither be pointed at anyone nor should compressed air be used to dust off clothing or any part of the body.

Improper use of grinding wheels is often the cause of accidents. Grinding wheels should be properly mounted and have a tool rest, a protective glass shield, and a guard covering most of the wheel. Goggles or a face shield should always be worn. Only the face of the wheel, never the side, should be used for grinding operations.

Gas welding rigs that use acetylene and oxygen should only be used by persons who are thoroughly trained in their operation. All welding equipment should be properly maintained. Protective caps should remain in place until cylinders are to be used. Proper turn-on and -off procedures should be used. The oxygen should never be used as a substitute for compressed air. Oil and grease should be kept from oxygen equipment since oxygen unites explosively with these materials.

Persons in an area where welding, either gas or arc, is taking place should not look at the welding operation, even for a moment, unless the eyes are protected.

Storerooms and Stockrooms

Good housekeeping should always be practiced in storerooms and stockrooms for a safe and efficient operation. All aisles and walkways should be kept clear of stock and waste materials. All personnel should know the location and operation of fire-extinguishing equipment. No equipment should be placed in front of the fire extinguishers.

All equipment and material should be properly stored. Stocked items should be properly arranged. Heavy items should be stored at lower locations.

Personnel working in storerooms should wear gloves and safety shoes. All crating material should be handled with care and properly disposed of. Hand trucks should not be overloaded.

Only authorized trained personnel should be allowed to use fork lifts, which should never be overloaded or operated at high speeds. Operators should always face in the direction of travel and be particularly careful when backing up.

Vehicle Driving

All drivers should learn to be safe and defensive drivers. A good driver checks and maintains the vehicle properly, uses the safety equipment, shows courtesy to other dirvers, signals well in advance of any directional change, takes pride in developing driving skills, and knows and observes all traffic regulations.

Every vehicle should be thoroughly checked before it is put on the road. The lights, horn, muffler, oil, gas, and water should be checked. In bad weather, the defroster, windshield wipers and snow chains, if necessary, should be checked.

Before the vehicle is started, the mirrors, seat belts, and seats should be properly adjusted. If flares, flags, or fire extinguishers are provided, they should be in good operating condition. If the load on a vehicle extends beyond the bed of the truck, it should have a flag.

All personnel should remember that when operating company vehicles they are representing the company. Proper driving habits lead not only to safer driving but also to better public relations. This is especially important for companies that are publicly supported.

Backing accidents can easily be avoided if proper precautions are taken. A physical check should be made, if doubt exists, to ensure that backing is safe. In tight situations, a guide behind the vehicle should be used. A vehicle should not be backed more than is absolutely necessary. Whenever possible, park in a position that will not later require backing.

Every vehicle should contain accident forms. If an accident should occur all drivers should know how to fill out the forms.

Drivers should always use extreme caution at intersections, even if traffic flow there is controlled by a light. Always be ready to yield the right-of-way if necessary.

Tailgating is a dangerous practice that should be avoided. A minimum of one car length for every 10 mph should be left between vehicles. If visibility is limited this distance should be increased.

When parked, be certain parking brakes are properly set. On inclines, wheels should be blocked.

Working in Manholes

Underground structures should never be entered until one is assured that the air is safe. Oxygen and gas detectors should be used to check for harmful gases and for sufficient oxygen. If the manhole should contain insufficient oxygen and/or dangerous gases, then the structure should be ventilated until the condition is corrected.

Fresh air should be pumped into deep structures and any structure that does not have adequate ventilation. If a blower is used, the discharge end of the hose

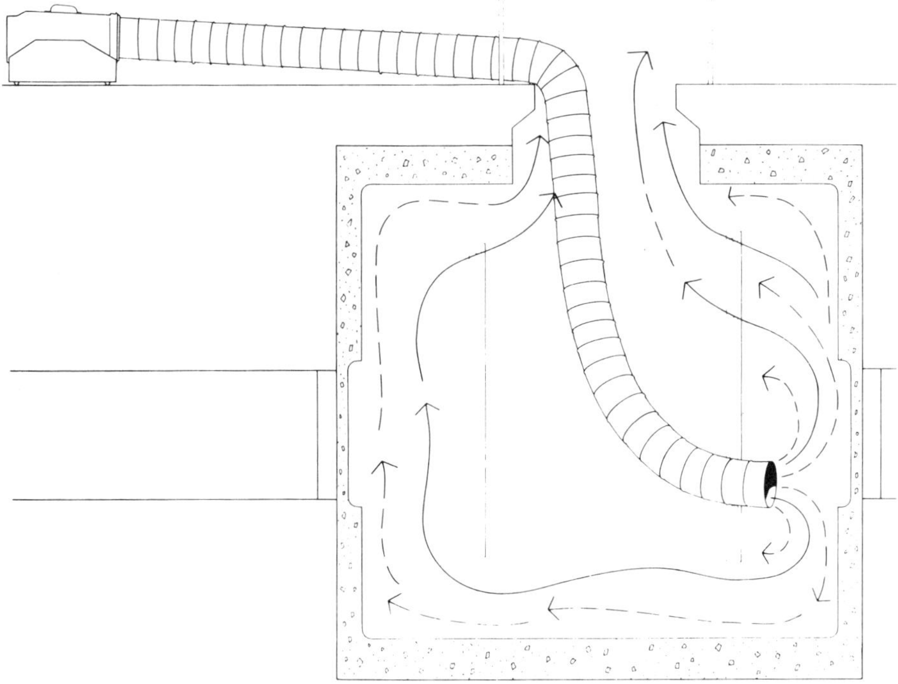

Fig. XIII - 1 Ventilation of a Manhole

Suggested Rules for Working Manholes and Confined Spaces

1. Test manhole for combustible gases and lack of oxygen.
2. Purge blower hose for a minimum of one min. above ground before placing in manhole.
3. Purge manhole for five min. before entering.
4. Allow no flame within 10 ft of manhole.
5. If duct plugs are removed while working in manhole, make additional tests and repreat #1-3.
6. If ventilating blower stops, get out of manhole immediately. Repeat #1-3.
7. Repeat testing procedures every 2 hr while work is required in the manhole.

should be placed near the bottom of the structure. This forces the old air up and out as shown in Fig. XIII-1.

The buddy system should always be used when working in manholes to allow one person to be stationed at the top of the manhole. This person can save steps by raising and lowering needed tools, and can also operate rescue equipment if something should happen to the other person.

Care should be exercised in removing manhole covers. The use of mechanical assists will prevent any possible foot or back injuries if properly used. (see Fig. XIII-3).

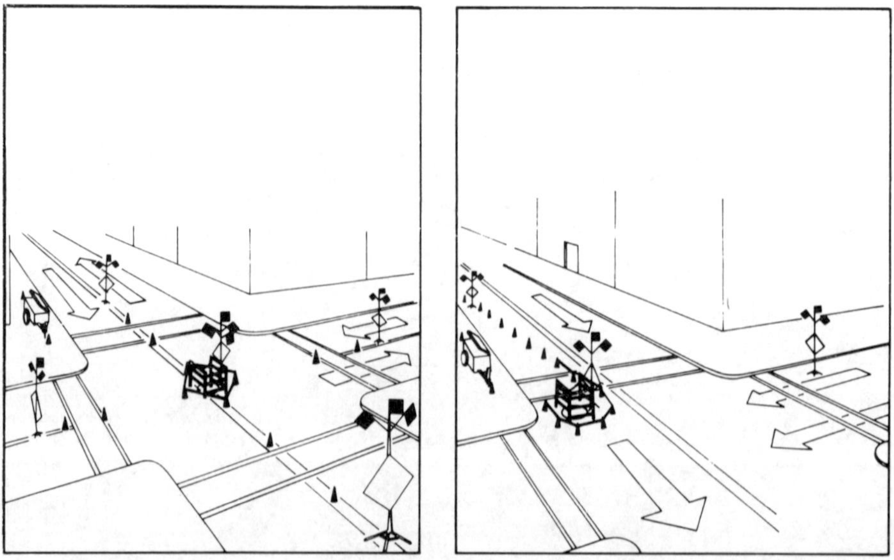

Cones should be positioned so as to effect a guidance path similar to that which is accomplished through striping to direct vehicles into a Left Turn Lane. If cones are spaced correctly with relation to speed of traffic they will produce the same effect on an approaching motorist as a solid line.

Speed—*mpg*	Distance—*ft*
up to 30	10 - 20
30 - 45	25 - 35
45 - 55	40 - 50
55	55 - 60

Fig. XIII - 2 Placement of Traffic Cones

Traffic should be carefully controlled when working in manholes in the street (see Fig. XIII-2). If possible, a vehicle can serve as a protective wall between the structure and the traffic. All barricades, lights, cones, signs, and other traffic equipment must meet state regulations (see Fig. XIII-3).

Guard rails should be placed around any open and unattended manholes. A toe board around the manhole should be used to prevent objects falling into the structure. Individuals in the manhole should also wear a hard hat for additional protection.

Fire Protection

All personnel should know what to do in case of a fire. If fires are dealt with quickly, they are less likely to cause damage. In order to deal with a fire

Table XIII-2

Types of Fire Classifications

Class	Material
A	Wood, paper, cloth
B	Gas, oil, grease, paint
C	Electrical

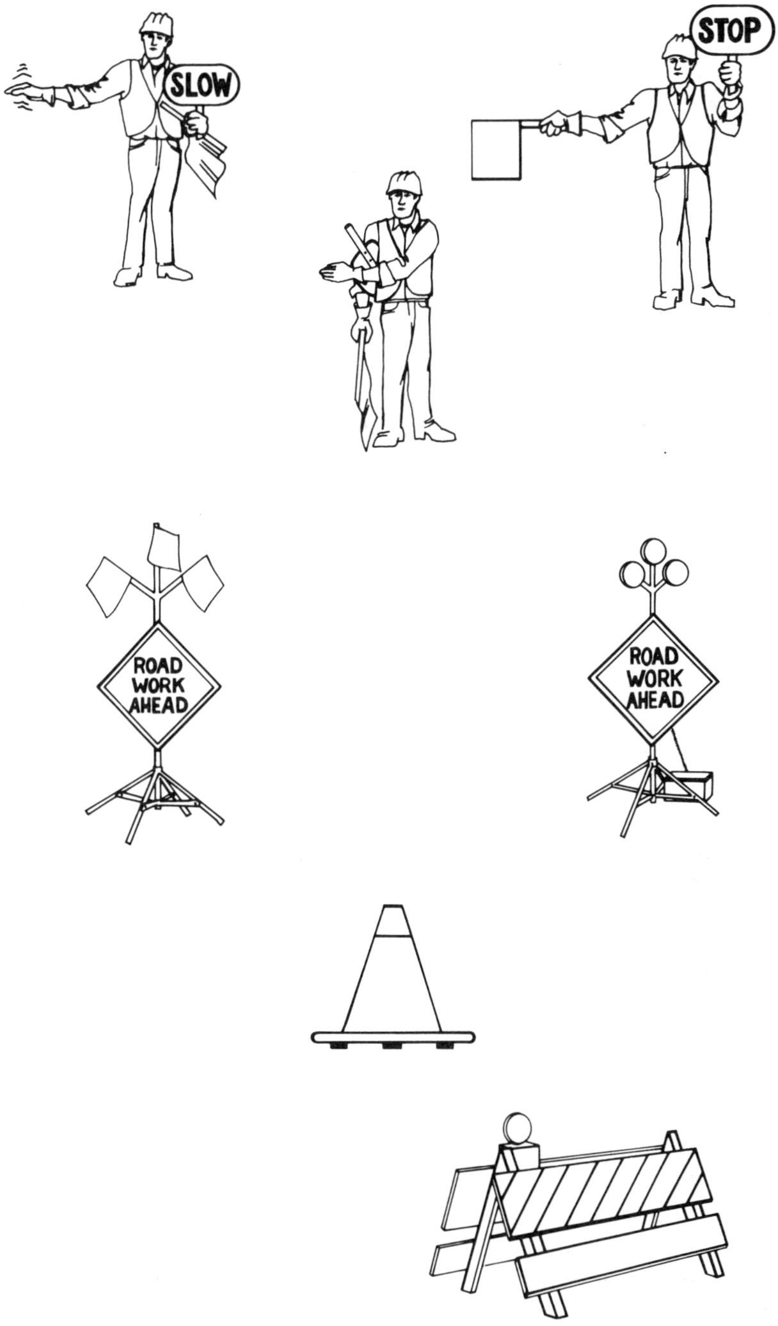

Fig. XIII - 3 Common Warning Signals

promptly, the proper type of fire extinguisher must be used.

Fires are classified into four classes. Class A fires are fires of ordinary combustible material, such as wood, clothing, or paper. Class B fires are of flammable liquids, greases, paints, or similar material. Class C fires are in or near electrical equipment. Class D fires are of combustible metals such as magnesium. Class D fires are rarely encountered and *will not* be covered here.

Water may be used only on Class A fires. Never direct a stream of water on a Class B or C fire. Most extinguishers are effective to some degree on Class A fires. Class B fires are fought with extinguishers issuing foam, carbon dioxide (CO_2), or dry chemicals. Use only CO_2 and dry-chemical extinguishers on Class C fires. If possible, turn off power to electrical fires. See Table XIII-2.

Fire extinguishers should be conspicuously located with signs or lights nearby to assist in their location. All personnel should know the type of fire for which extinguishers are meant and their proper use.

Most companies have a policy of inspecting all fire equipment on a regular basis, usually quarterly or semi-annually. However, all extinguishers should be refilled promptly after their use.

All exits should be marked. Flammable storage areas should be marked with signs and smoking should not be permitted.

All personnel should know the phone number of the local fire station, which should be posted conspicuously to ensure prompt response by the local department.

The best way to fight fires is to prevent them from occurring. Fires are often caused by careless cigarette disposal, oily-rag storage, electric arcs and combustible gases.

Trench Construction

Personal safety equipment is required when working in a trench. All persons must wear a hard hat. Personnel who work with jackhammers or who are exposed to flying particles should wear adequate eye and foot protection.

All material excavated from the trench must be stored at least 2 ft from the trench. If mechanical equipment is being used to excavate the trench then workers in the trench should face the machinery.

Trenches 5 ft or more in depth must be shored unless the bank is cut back at an acceptable angle. Trenches should be braced from the top first. Trenches in soft soil may require extensive shoring and bracing even at shallow depths. When employees are required to be in trenches 4 ft deep or more, an adequate means of exit, such as a ladder or steps, must be provided and located no more than 25 ft away from them.

Equipment Operation

Only authorized operators should run large pieces of equipment such as dozers, backhoes, and cranes. It is their responsibility to see that nearby personnel are clear of the machine, that the safe load limit is not surpassed, that the equipment is not operated near electrical lines, and that the equipment is left in a safe condition. Operators should always wear hard hats and use all required safety devices.

Pipe

All pipe should be carefully handled. If several people are moving pipe by hand, care must be taken to see that they work together. One person should give all signals or commands.

If skids are used, they should be adequate for the pipe and secured firmly. Snubbing ropes must also be adequate and in good condition. No one should stand between the skids while the pipe is lowered. Personnel operating the ropes must wear gloves to prevent rope burns. If necessary, wood chocks should be used to control the pipe.

If pipe is handled mechanically, such as by a crane, only one person should give directions to the operator. The pipe should be carefully rigged, and no one should ever be allowed beneath a raised pipe.

If pipe is stored, either in a yard or at the job site, it should be adequately blocked and stacked to prevent it from rolling or falling. All pipe should be stored according to lengths and sizes. Threaded pipe must be carefully handled and stored to prevent damage to the threads. Pipe storage areas should be maintained in a clean and orderly manner to prevent accidents. Aisles should always remain clear.

Lead melting and caulking are hazardous jobs that should only be done by persons properly trained in the correct methods. No distractions or unsafe conditions should be present during these operations. All persons working near the hot material should be alert and careful. Only the proper tools should be used, including personal protective equipment such as gloves and face shield. Caulking compound should be handled with the same precautions as melted lead.

Barricades and Warning Signs

Barricades, warning signs, traffic cones and flashing lights are frequently used in operations of a water utility. These devices should be placed so as to warn traffic adequately of the work or hazard ahead. If necessary, flagpersons should be used to divert or slow traffic. All persons working near the traffic should also wear bright reflective vests.

All barricades and signs must now meet certain OSHA standards for color and placement. Care should be taken that these standards are being met.

Chlorine

Chlorine is used quite extensively by water utilities. This chemical must be handled, stored, and used properly if no accidents are to occur. But if an accident should occur, knowledge of the proper steps to take may prevent the accident from being serious or fatal.

Everyone who works around chlorine should know the characteristics of the gas and be able to recognize a gas or liquid leak.

As a gas, chlorine is greenish—yellow in color and is about 2½ times as heavy as air. As a liquid, chlorine is amber colored and about 1½ times as heavy as water. Unconfined liquid chlorine will rapidly vaporize to gas. Chlorine has a pungent, biting smell.

Chlorine is chiefly a respiratory irritant, and, depending on the concentration and duration of exposure, may cause irritation of the skin, skin burns, irritation of mucous membranes and eye burns. Unnecessary exposures to chlorine should therefore be avoided.

Chlorine is very corrosive in the presence of water. Because of this, chlorine cylinder leaks should never be sprayed with water since that will only increase the leak.

All personnel should practice the proper use of gas masks. In case of a gas leak, personnel without a mask should immediately leave the area. Panic should be avoided. The mouth should be kept closed, and coughing or deep breathing should be avoided. The head should be held as high as possible since chlorine is heavier than air and will settle. Ventilation equipment should be used to remove the chlorine from the air.

Calcium hypochlorite is a dry-powder chlorine that is commonly used in chlorination. It is corrosive in the presence of moisture, it will react with organic material, and it will support combustion. Calcium hypochlorite will cause respiratory irritation, burn skin, and damage mucous membranes and eyes. If ingested, it is a poison. Therefore, it is necessary that only properly trained personnel handle and use this product. Personal safety equipment, including full dress, should always be worn.

Sodium hypochlorite is a strong, corrosive acid. It may cause the same injuries as other forms of chlorine. It too should only be handled by trained personnel wearing protective equipment.

Other Chemicals

Occasionally personnel will use other chemicals in water distribution. These include flourides, lime, soda ash, sodium hydroxide, and sodium chlorite. These chemicals should only be handled by trained persons. Safety equipment, including goggles, face shields, rubber aprons, gloves, and boots should be worn where appropriate.

First-aid instructions should always be posted where chemicals are handled and stored. Warning signs should be posted to prevent untrained persons from handling the chemicals. Showers and eyebaths should be provided where these chemicals are handled.

Epilogue

Every employee must take responsibility for the safe performance of required duties. This attitude should continue after hours in the person's home. An injury at home is just as painful and costly as one at work. In fact, the majority of disabling injuries do occur at home.

Everyone should try to use the same attitudes at home against unsafe acts and unsafe conditions as they would at work. Safety is a full-time job. *THINK SAFETY!*

Appendix — First Aid

Operators should be trained in first aid, but even without this training, there are some general rules to follow. See Table XIII-3.

One should not move a seriously injured person unless it is absolutely necessary.

Fractured arms and *legs* should be tied to splints before moving. If no splint material is available, a fractured leg can be tied to the other leg. However, a good rule to follow is not to move the patient unless there is a danger of additional injury.

Act fast if there is *severe bleeding* or if the victim has *stopped breathing*. You may have a minute or less to save a life. In all cases where breathing has stopped, apply artificial respiration. Bleeding should be stopped by applying pressure on a sterile bandage over the wound. If this doesn't work, pressure should be applied with the palm of the hand on an artery pressure point between wound and main body. These points are on the upper inner thighs and upper inner arms. No stimulants should be given to an injured person.

A victim of a serious injury should be made comfortable and the clothing loosened. He or she should be kept quiet and warm, preferably lying down with the head down. Liquids should be given sparingly, a sip at a time.

Exceptions to the preceding procedures would be a *head injury,* where the head and shoulders may be raised, or for *heat strokes,* where the head is raised and the body is cooled quickly by pouring water over the victim or by wrapping the victim in wet towels. Also, for heat stroke three or four glasses of cold water may be given containing ½ teaspoonful of salt, spaced about 15 min apart.

Cuts and *wounds* should be washed with soap and water. Wash away from the wound, don't scrub. *Puncture wounds* should be encouraged to bleed by squeezing. Then they, like chemical and heat burns, should either be immersed in water or rinsed with running water then covered with a sterile bandage. Don't pull clothing off a burn or wound, but cut the clothing away.

For *electric shock* remove the source of current as quickly as possible; every second counts. Pull the main switch, disconnect the plug, or pull the wire off the victim by using a dry non-metallic pole or broom.

Animal bites are treated as other wounds before taking the patient to a doctor except that the animal involved should be captured and identified if possible.

If the case of a *snake bite* have the victim lie down and stay quiet. Tie a constricting bandage, necktie, shoestring, or the like, several inches above the bite, just tight enough to dent the skin, retarding the lymph flow in the lymphatic system. Take the person immediately to the doctor, and keep the limb cool.

Chlorine detoxification.

The operator should

1. Remove victim from gas area into a warm room.
2. If breathing has stopped immediately start artificial respiration according to instructions described later in this section.
3. Promptly remove clothing contaminated with liquid chlorine or highly chlorinated water.
4. Place victim on his back with a folded coat or blanket under the shoulders causing his head to fall well back, maintaining air passage to lungs.

Blankets should be placed under and over victim. It is essential to keep the victim warm, quiet, and restful.
5. Call a physician immediately after taking steps 1-4.

In addition there are the following procedures.
1. A mixture of carbon dioxide and oxygen, with no more than 7 per cent carbon dioxide, may be given. This mixture can be administered for periods of 2 min followed by 2-min rest periods for no longer than 30 min. Instructions of the gas apparatus supplier should be followed carefully. If carbon dioxide and oxygen mixture is not readily available, then oxygen alone or fresh-air "Air Pack" may be used.
2. Milk may be given in mild cases as a relief from throat irritation.
3. When the victim's eyes are irritated by chlorine, they should be washed repeatedly with water. Each chlorination station should have an eye wash. In severe cases of eye contamination due to chlorine, the procedure should be to irrigate for 15 min, and rest for 10 min. A routine of 5-min irrigation and 10-min rest should then be followed for 1 hr. Prompt action is absolutely essential to protect eyesight.
4. Areas of the skin that have been splashed with liquid chlorine or chlorinated water should be repeatedly washed with water. After thorough washing, any burned areas shoud be covered with a sterile dressing and bandaged snugly unless blisters are apparent; in which case they are bandaged loosely.

If facilities are available, it is generally recommended that the victim be removed to a hospital as soon as possible, unless recovery from chlorine exposure is prompt and the exposure mild.

Before moving any seriously injured person call for medical assistance or advice. If possible, call for an ambulance to take the injured person to a hospital or physician. If transporting the injured person call ahead to the doctor either by telephone by through the mobile radio and dispatcher.

It is important to remember that in administering any first aid, you must remain calm, so as not to excite the injured person.

Artificial Respiration

No time should be lost in starting - delay can be fatal.

Preliminary operations.
The person giving artificial respiration should
1. Clear victim's mouth of any obstructions.
2. Lay casualty victim on his back.
3. Elevate shoulders of victim with a folded coat, blanket, etc. so his head falls well back. This maintains a clear air passage to his lungs.

Revised Sylvester method
1. Place victim's head between knees and grasp his arms at the wrist.
2. Cross arms over the lower half of the breastbone and rocking forward, press firmly downwards (about 20-lb pressure), forcing air from lungs of victim.
3. Release the pressure by rocking back and pulling his arms upward, out-

Table XIII-3
Symptoms and Treatment

ASPHYXIATION	Unconsciousness. Lips, earlobes blue, breathing stopped	Remove to fresh air. Give rescue breathing (see text).
ARTERIES BLEEDING	Bright red blood spurting from wound	Apply direct pressure with sterile compress. Press at pressure point (see text).
BLEEDING—FROM VEINS	Dark red blood flowing steadily	Apply sterile bandage firmly over wound. Do not break blood clot.
BLEEDING—INTERNAL	Faintness, pale face, thirst, sighing, weak rapid pulse	Lay down, keep warm and quiet.
BURNS	Redness of skin, blisters, charring, in order of severity	Apply cold compress. Deep burns, apply sterile bandage only. Flush chemical burns with water.
DROWNING	Unconsciousness, no breathing	Press stomach to remove water. Clean mouth. Loosen clothing. Give rescue breathing (see artificial respiration, text).
DISLOCATIONS	Lengthened or shortened limb, unusual protrusion at joint, unusual angle at joint	Support along lines of deformity. Do not attempt to reposition.
ELECTRIC SHOCK	Unconsciousness. Breathing stopped, burn at point of contact	Shut off electricity or insulate yourself, then rescue victim. Give rescue breathing (see text).
FAINTING	Face pale. Eyelids droopy. Cold sweat. Weak, rapid pulse	Lay down. Loosen clothing. Keep warm and quiet.
FRACTURES	Pain, swelling. Unable to move part beyond	Support limb above and below fracture. Apply well-padded splints if injured must be moved.
FRACTURES—COMPOUND	Open wound, bleeding. Bone protruding	Bandage wound. Apply splints in line of deformity if injured must be moved. Do not straighten.
FRACTURE—SKULL	Possible bleeding from eyes, nose, mouth. Blood or serum from ears. Swelling of head wound. Depression in skull	Elevate head. Lay bandage on injury. Keep warm. Never give stimulants.
GAS POISONING	Yawning, giddiness, fluttering heart, breathing stopped	Remove to fresh air. Give rescue breathing (see text).
HEAT EXHAUSTION (exposure to heat or sun)	Cold, wet skin, dizziness, headache, nausea	Lay down. Loosen clothing. Keep cool.
RUPTURE	Sharp stinging pain. Faint feeling, sick at stomach. Lump in groin	Place injured on his back, elevate knees. Apply cold pack to lump.
SHOCK	Face pale, eyelids droopy. Skin cold and clammy. Partial or totally unconsciousness	Lay down. Loosen clothing. Keep warm and quiet.
SNAKE BITE	Pain, swelling, pinpoint wounds	Apply tourniquet 3 inches above wound just denting skin. Apply cold to area. If alone, walk slowly for help. Be calm. Don't squeeze or cut wound.
SUNSTROKE (exposure to heat or sun)	Hot, dry skin, fast pulse, dizziness, headache, nausea	Lay down. Loosen clothing. Apply cold packs to body and head.
WOUNDS	Break in skin, incised, lacerated, punctured	Cover with clean sterile bandage. Call doctor for all puncture wounds.

ward, and backward. This extends the chest walls and draws air into the victim's lungs.
4. Repeat cycle 12 - 15 times per minute.

Oral Resuscitation

The person giving first-aid should

1. Kneel beside victim's head and tilt head back, pulling the jaw up.
2. Pinch victim's nose, open mouth wide and blow into victim's mouth strongly enough to cause the victim's chest to rise.
3. Remove mouth. Victim's chest should fall.
4. Repeat cycle 12 - 15 times per minute.

Artificial respiration must be continued until natural breathing is restored or a doctor says to stop.

Cafe Coronary

A new method has been developed by Dr. Henry Heimlich of Cincinnati to dislodge food in the windpipe of a choking person, a condition commonly called a *cafe coronary*. It is estimated that more than 100 lives were saved in the first few months that this procedure was known.

To use this method, one should stand behind the victim and put both arms around him. Let victim's head, arms, and upper torso hang forward. The person performing the aid should grab his fist with his other hand and place it against victim's abdomen slightly above the navel and below the rib cage, pressing up rapidly against abdomen. This forces the diaphragm up and compresses the lungs. Hopefully, the food will pop out like a cork from a champagne bottle.

If the victim is too heavy to hold, lay him on his back and sit on his hips. With the heel of the left hand pressing against the back of right hand, the person giving aid should push forward into the victim's abdomen just above the belt.

A person choking on something while alone, should use the technique on himself by pressing fist rapidly up against abdomen.

•••••

All the cases cited here require professional medical attention. The task of the person giving aid is to perform that which is necessary before the doctor arrives or the patient can be transported to a medical facility.

Chapter XIV
Chlorine Handling and Safety

This chapter teaches safe, approved methods of handling chlorine. Chlorine deserves respectful care when being handled. The water system operator has the responsibility to be trained in the proper use of chlorine and care of equipment to provide the needed disinfection while ensuring that the chlorine does not become a health menace to himself or others.

Chlorine Hazards

Chlorine can be detected by smell, even in very small dosages. The least detectable amount of chlorine in the atmosphere is about 3½ ppm, and when this occurs the operator should be alerted to potential hazards, such as leaks or faulty equipment. The maximum amount that can be inhaled for 1 hr without serious effects is about 4 ppm. At 15 ppm, chlorine will cause irritation of the throat; at 30 ppm, it will cause serious coughing spells; and at 40 - 60 ppm, it is extremely dangerous even for a ½-hr exposure. A few breaths of air containing 1000 ppm can kill a man.

Handling 150-lb Cylinders

When unloading full 150-lb cylinders, they should not be dropped from the truck and they should be prevented from falling over or falling against each other.

When moving cylinders, use of a hand truck with a clamping arrangement or safety chain at least two thirds the way up the cylinder is recommended.

An experienced operator can safely move a cylinder by rolling it on its bottom edge. When rolling a cylinder in this manner, it should not get out of control and fall; its protective bonnet should not turn loose.

When chlorine cylinders are handled, the protective bonnet should be in position. It should be removed only when the cylinder is connected to the chlorinating system. To lift a cylinder when an elevator is not available, use a crane or hoist with a special cradle. Chains, rope slings, or magnetic devices should never be used for lifting. Never lift a cylinder by the valve-protecting bonnet.

Storage: Empty cylinders should be stored in one area and full cylinders in another area. Cylinders must be stored in an upright position and restrained from toppling by using a safety chain anchored to the wall with a snap hook placed around them. The storage area should be well ventilated.

Valve: The 150-lb cylinders are equipped with a single Chlorine Inst. standard

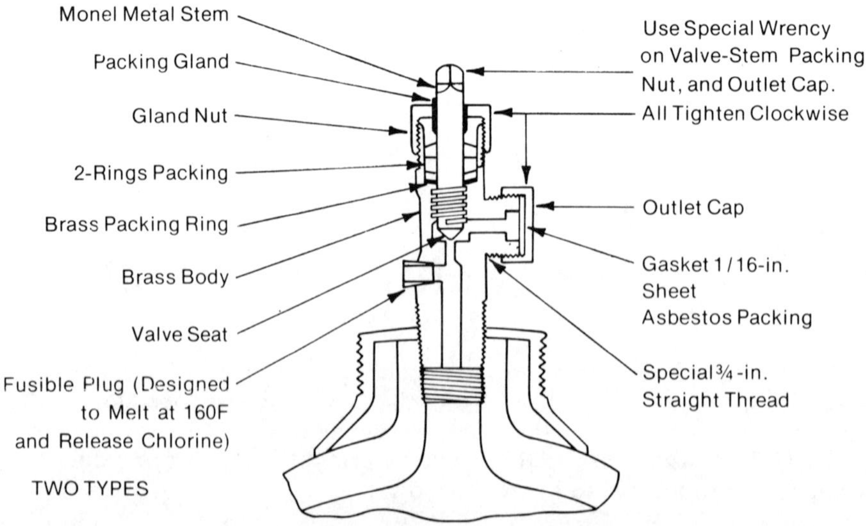

Fig. XIV-1. Cross Section of Valve in a 150-lb Cylinder

cylinder valve and a fusible plug (Fig. XIV-1). Valves are tested and reconditioned or renewed by the manufacturer after each trip. The fusible plug is located on the side of the valve opposite the outlet and will melt at a temperature of 160F. Its purpose is to release chlorine if pressure becomes excessive because of fire or overheating.

Use

Connecting cylinders by clamp and adapter. One should connect a cylinder to the chlorinating system by

1. Securing the cylinder to a building column or solid, upright support.
2. Removing the protective bonnet. If the cylinder has been exposed to the weather for a long time, the threads at the base of the bonnet may have corroded, in which case a few smart raps on opposite sides of the bonnet will loosen it so it can be unscrewed easily.
3. Removing the brass outlet cap and any foreign matter that may be in the valve outlet recess. A 2-in. nail can be used to clear out the old washer.
4. Placing a new washer in the outlet recess.
5. Placing the clamp over the valve. The adapter is inserted in the outlet recess, and then, with the adapter fitted into the clamp slot, the clamp screw is tightened. The end of the adapter should seat firmly against the washer. A cylinder connected to the coil is shown in Fig. XIV-2.

At no time should anyone reuse washers. At no time should anyone lift a cylinder by its bonnet.

The protective hood on the 150-lb cylinders is screwed onto a threaded neck ring. Despite its appearance, the neck ring is actually not part of the cylinder, and is often not securely attached to the cylinder. *The hood should always be in place except when the cylinder is in use.*

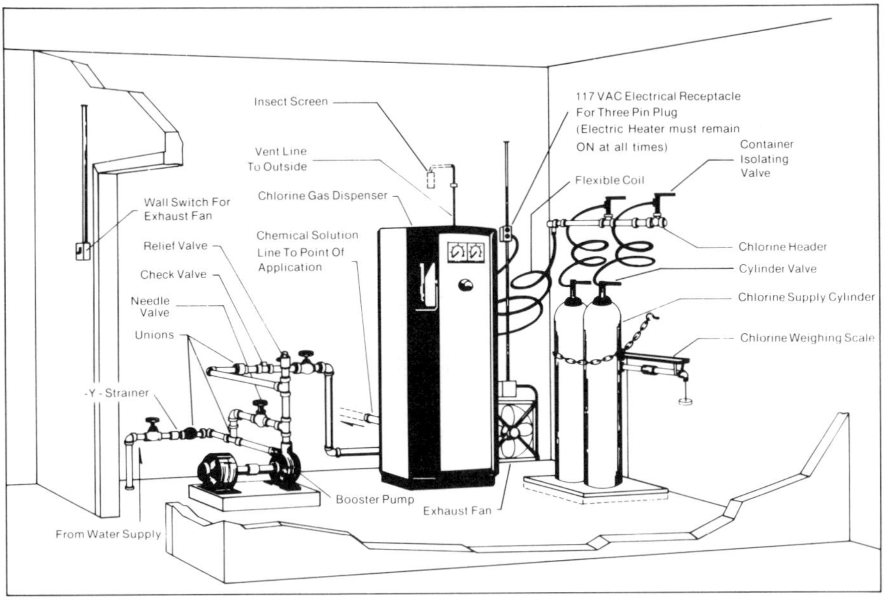

Fig. XIV-2. Typical Gas-Dispenser Installation
A- overview

Connecting cylinder - using threaded connections: The operator making connections with piping having threaded couplings instead of the yoke type should *use two wrenches:* a large spanner is used for the coupling, and a small spanner is used for holding the squared area of the pipe itself at the coupling. Use the flat spanner or box-end wrenches supplied by the manufacturer for all chlorine-cylinder pipe connections.

Weighing the cylinder. The only reliable method of determining the contents of a cylinder is to weigh it and check its weight against the tare stamped on the cylinder shoulder. The pressure in a cylinder depends upon the temperature and does not indicate the amount of chlorine present.

One-Ton Cylinders

Storage and handling. *One-ton cylinders must be moved by an approved lifting bar and hoist.*

General storage conditions are the same for both one-ton and 150-lb cylinders, except that one-ton cylinders are always handled and stored in a horizontal position. A one-ton container has two valves very similar to the one on the 150-lb container; in fact the only difference is that these valves (Fig. XIV-3) do not have a fusible plug. Instead, the one-ton cylinder has three separate fusible plugs that will melt at 160F and discharge the chlorine from within. The delivery rate of chlorine from a one-ton container will depend on the temperature of the liquid in the container, but an average flow is about 20 lb/hr of chlorine gas.

The one-ton cylinder is set on the scales with the two discharge valves one above the other. The top valve will discharge chlorine gas. The bottom valve will discharge chlorine liquid. (See Fig. XIV-4.)

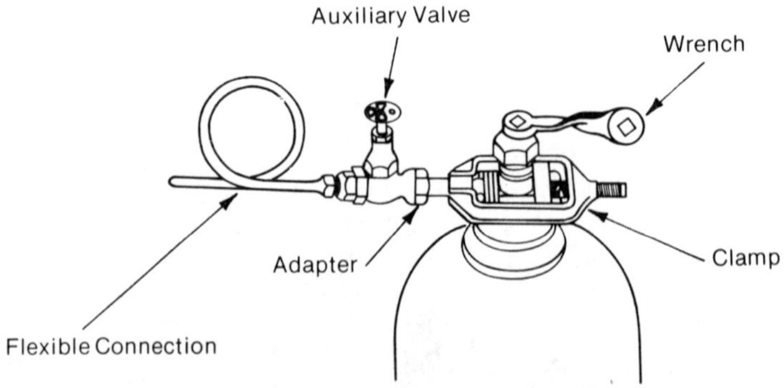

Fig. XIV-2. Typical Gas-Dispenser Installation
B- yoke installation

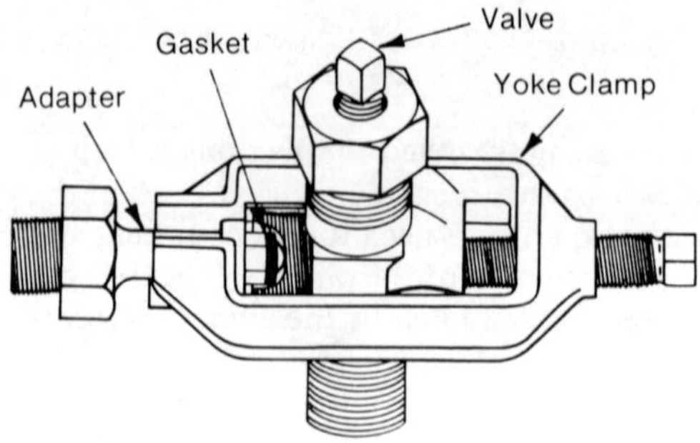

Fig. XIV-2. Typical Gas-Dispenser Installation
C- yoke in detail

The procedure for connecting and testing valves and feed pipes of one-ton cylinders to header or chlorinator is the same as described for the 150-lb cylinders and should be followed accordingly.

Valves

Opening the valves. Place the wrench provided by the manufacturer on the valve stem and, standing behind the cylinder valve outlet and grasping the valve firmly with the left hand, hit the wrench a sharp blow in a counterclockwise direction with the palm of the right hand. There should be no pulling or tugging of the wrench since this may bend the stem, causing it to stick, and, once opened, the valve may not close properly. *Under no circumstances use an ill-fitting manufacturer's wrench or pipe wrench* since these wrenches will round the corners of the squared end of the valve stem.

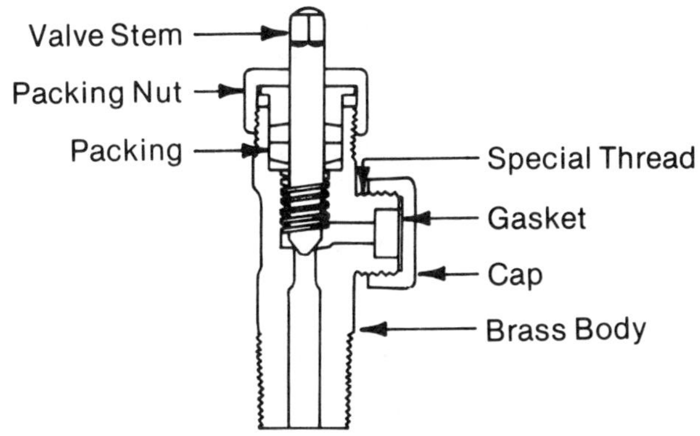

Fig. XIV-3. Cross Section of Ton-Container Valve

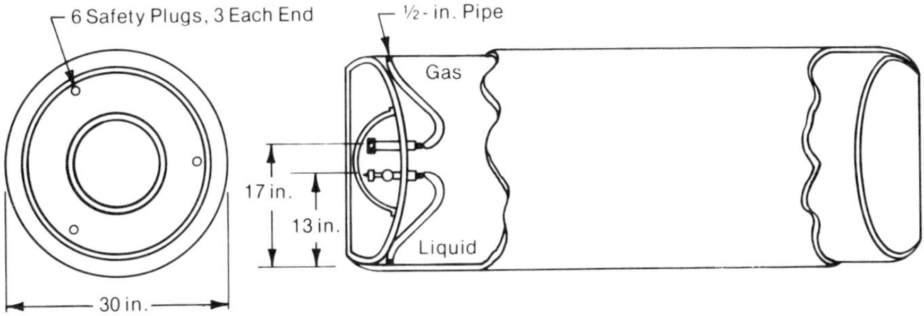

Fig. XIV-4. Cross Section of Ton Container

The use of large wrenches for opening stubborn valves should be avoided because there is danger of bending the valve stem or breaking it. To open a stubborn valve, the normal opening procedure is followed while using a small block of wood held in the palm of the hand when striking the wrench. The valve should be opened slowly. One complete turn permits maximum discharge. The valve should only be turned as far as it will go without removing the wrench for another bite. The wrench should be left on the valve to allow for easy and rapid shutoff in an emergency.

Closing the valve. When all the chlorine has been discharged from the cylinder and after the pressure in the cylinder reaches zero the valve is closed. Closing should be done with the wrench provided while the valve is grasped in one hand and the wrench is tapped in a clockwise direction with the palm of the other. If the valve does not close tightly on the first trial, it should be opened and closed lightly several times until proper seating is obtained. *A hammer or other tool should never be used to close the cylinder valve tightly.* The outlet cap should be replaced immediately upon disconnecting the cylinder so that the valve parts will be protected from the moisture in the air.

The protective bonnet should be emplaced as soon as the cylinder is disconnected. The outlet cap of each valve is fitted with a gasket that is designed to fit against the valve outlet face. If a valve leaks slightly after closing, the leak may be stopped by drawing up the valve cap tightly.

Valves should never be operated in such a manner as to isolate chlorine gas or liquid in a line. In some installations using one-ton cylinders there is a short length of tubing running between cylinder(s) and evaporator(s). There is a valve on the cylinder and one on the evaporator line. The tubing between the two is full of liquid chlorine during normal operation. If for some reason the valve on the evaporator line is closed and then the cylinder valve is shut, the tubing is left full of liquid chlorine. A small increase in temperature will cause a considerable increase in gas pressure. With the tubing full of chlorine and closed at both ends, there is no room for gas expansion, and a potentially dangerous situation exists.

The safe procedure is to close the cylinder valve first and then allow sufficient time for the chlorine in the tubing to be exhausted by the evaporator before closing the evaporator inlet valve. This is the procedure followed when changing cylinders to allow the chlorine to be drained from the ''pigtail'' tubes. Closing the evaporator outlet valve is dangerous unless the evaporator is shut off and cooled.

Valve stiffness. Valves are carefully checked before leaving the manufacturer's plant, but occasionally a valve may be stiff to turn or difficult to shut off tightly. This condition may be caused by a small piece of scale or other foreign matter at the valve seat. Sometimes the valve can be freed by opening and shutting a few times (with the outlet cap in place and by tapping the body of the valve). Actually, once the cylinder is hooked up, inability to shut the valve completely is not important and contents can be withdrawn until empty. After the cylinder has been discharged (and only gas remains in it) the outlet cap with a good fiber gasket will effectively stop chlorine from escaping.

Starting Up Chlorinator Using Two or More Cylinders

Water to the chlorinator injector should be turned on, making the necessary adjustments to the machine for operation when the gas is turned on.

The valve on the cylinder farthest from the chlorinator should be cracked open until approximately 40 psi of pressure is recorded on the pressure dial of the chlorinator.

The cylinder valve should then be shut to test the system for leaks. If there are no leaks, the procedure just described is repeated consecutively for each cylinder until all cylinders have been tested. If a gas leak develops during the checkout of pipes and cylinders, then the operator should leave the chlorine room and wait for the ventilating system to clear the gas from the room. Since the chlorine cylinder was shut off earlier, the chlorinator needs only the use of the gas in the piping.

If no pressure is recorded when the valve has been opened, it may be that the feed pipe is plugged or other operating problems may exist. *The valve should be closed immediately. No attempt to repair or make adjustments to any part of the chlorine system should be attempted until the chlorine gas trapped in the feed pipes from cylinder to chlorinator has been released.* To do this, the operator should first put on a self-contained air pack with full face piece. He should then loosen one of the feed-pipe connections and allow the trapped gas to escape slowly from the piping. Only after the room has been cleared of all escaping gas should the necessary repairs or adjustments be made.

Operating Safety Instructions

1. One should always turn on the chlorine exhaust fan when entering the chlorine room, whether for a routine check or for servicing or repairing leaking equipment even while wearing a mask.
2. Whenever the concentration of the chlorine gas in the air is unknown, a self-contained air pack must be worn.
3. No one should start up or operate a chlorinator or turn on a chlorine cylinder unless adequate protective equipment (air-pack respirator) is on hand in the chlorine room area.
4. A self-contained air pack must be worn where it is necessary to locate and stop small leaks in the piping or when making any repairs or adjustments to leaking equipment.
5. All persons using the gas-protective equipment must be trained in its use and maintenance.
6. A self-contained air breathing unit must not be used unless the air cylinder is fully charged. Air cylinders must be completely recharged after each use.
7. If a leak occurs, people occupying a room should vacate it and reenter the room only after donning self-contained air breathing units—and then only to shut off the chlorine supply.
8. When a leak occurs in a chlorinator room, all doors leading into any other part of the building should be sealed. Only doors to the outside should be opened.
9. Water should never be applied to a chlorine leak because a water and chlorine mixture is highly corrosive and dangerous.
10. Plastic-coated gloves should be worn when changing cylinders.
11. Whenever possible two men should be present when changing chlorine cylinders or whenever chlorine piping is broken apart.
12. One should never lift a cylinder (150 lb) onto the scales. A ramp should be provided.
13. When connecting the piping from the cylinders to the chlorinator, or whenever it is necessary to break a connection and recouple it, one should always use a new lead or fiber washer.
14. The yoke, adapter, and clamp should be cleaned with a wire brush.
15. The yoke adapter should be inspected for rounded or worn edges at the contact point of the adapter clamp.
16. The piping from the cylinders to the header located on the wall or to the chlorinator must have an inverted loop of not less than 10-in. diameter. The loop acts as a flexible coupling.
17. Flexible-metal chlorine-gas piping should be replaced immediately if bent or twisted during cylinder changeover.
18. Piping in chlorination plants should be color coded and identified.
19. Petroleum-base fluids should not be used in contact with chlorine.
20. Each chlorination station should have a readily accessible eye wash and shower.

Feed-pipe support. The feed pipe or flexible coil from the header to the cylinder should be supported while the empty cylinder is being replaced by a full one. For example, support it on another cylinder, use a hook, or block it from the floor. This will prevent any kinking or weak spots from developing in the pipe.

If the pipeline is disconnected for any length of time, there is a danger of moisture forming in the line. The open end of the pipe should be plugged using a cloth, and the auxiliary valve should be shut.

Emergency Breathing Units

There are many different types of filter and self-contained breathing units available. Air-pack and canister units are the most commonly used in the water supply field around chlorine gas. Each person who may be called upon to use these "gas masks" should become familiar with their use.

The 15- and 30-min fresh air pack. One should check the pressure gauge on the cylinder periodically to see that it reads "full" at all times while stored. To deploy, the operator should open valve one complete turn on back of 15-min air pack. On 30-min air pack the "red by-pass" valve located by the harness air gauge must be closed and the "yellow demand" valve locked open at all times.

Putting unit on. The mounting, wearing, and use of breathing units should be practiced by all operators working around chlorine, until they are quick and comfortable in the deploying of these units. Careful following of directions is imperative.

The face mask should be tested for air tightness by placing a hand over the hose and taking a deep breath. If the mast draws in around face, air tightness has been obtained. *A face mask may not seal over a bearded face unless the beard has been heavily greased.*

The air pack is now working.

A low-air-pressure alarm on the 30-min air pack is activated when approximately 500 psi of air are left in cylinder. The operator should leave the contaminated area for a fresh cylinder of air once the alarm sounds.

To remove unit: After the mask is removed the air cylinder valve is closed and the system purged by opening the red bypass valve and closing it. *No unit should be stored without extending all straps to their limits and without refilling air cylinder.*

Canister-type gas mask. Many operators of water systems are using canister-type gas masks for protection when a chlorine leak occurs. *This type of mask provides only limited protection against chlorine gas, even under ideal conditions.* However, if a canister-type mask is all that is available for protection, it can be worn with reasonable safety if maintained properly and its limitations are understood.

The most important fact to remember is that a canister type mask will give no protection when the oxygen in the air is too low to support life, regardless of the condition of the mask.

The canister. The canister filters out chlorine gas in the air by adsorption. The number of times it can be used before it becomes exhausted is difficult to determine; it depends on how much chlorine is present in the atmosphere during each use. A new canister will protect for 10 min in a 2 per cent concentration of chlorine gas to air. After that, it is useless.

When a canister is in use, the flap over the intake part on the bottom of the canister is pulled aside to allow the air in the room to enter and pass through to the face piece. This flap must be replaced when the canister is not in use. If it is not replaced, the activated-carbon filter loses its adsorption qualities and will provide little or no protection the next time it is used. The shelf life of an unused canister is not really known, but the canister should be discarded after a year. *The only sure way to get the maximum protection from a canister is to use it only once and then throw it away.*

Air capsule. A new self-contained breathing apparatus for emergency escape from smoke- or gas-filled environments that was developed for the Navy is now available for civilian use. Simple to use, the device can be donned and activated in fewer than 10 s, and protects the user's entire head area with a transparent fire-resistant hood similar in appearance to a heavy plastic bag. This hood may be worn over a hard hat. A rechargeable air reservoir dispenses 5 min of continuous cool, breathing air, providing ample time to slip the device on and make an escape, or to make emergency adjustments of valves or other devices.

The hood maintains a positive air pressure at all times, thus keeping out toxic gases or smoke even if the wearer has eye glasses or a bearded face. A constant-flow regulator ensures 5 min of continuous air regardless of breathing rate. Because the unit does not make use of a mouth piece or nose clip, breathing is normal at all times. The wearer may also communicate easily while retaining optimum visibility. A combination elastic-drawstring-neckband comfortably fits around the neck.

The air reservoir is a double coil of stainless-steel tubing containing 5 cu ft (at 14.7 psi) of breathing air that has been pressurized to 5000 psi. To use, one unfolds the plastic hood and pulls the start ring at end of air reservoir, releasing an even flow of cool air into the hood; the hood is slipped over the head with the reservoir at the back of the neck and the neck drawstring is adjusted. Because it is quick and simple to don and activate, this breathing unit is very suitable for emergency use in an area such as a manhole where toxic gases or lack of oxygen may be unexpectedly encountered.

Leak Safety Procedures

A person who smells chlorine should

1. Procure and adjust gas mask for immediate use.
   Note: Gas masks should be located outside of the chlorine room beside the door leading to the chlorine room.
2. Turn on exhaust fan of chlorine room
3. Shut off cylinders; if leak is large, or if the equipment has been leaking for some time, the room is full of gas.
   Note: No tests should be made on equipment for leaks until room is cleared. No repairs should be attempted by an individual working alone.
4. Notify supervisor immediately.

Leak Detection

To test for leak the operator should

1. Wait for room to clear of all chlorine.
2. Crack open cylinder farthest from chlorinator until gauge shows approximately 40 lb.
3. Shut off cylinder and test system to find leak(s).
4. Use bottle of concentrated ammonia, swabbing near joints, piping or valves suspected; white fumes (ammonium chloride) will indicate the location of a chlorine leak. It may be necessary to repeat this action several times to find all leakage.
5. Mark leak off clearly once detected.
6. Shut off gas supply and keep supply shut off until leak is repaired.

Types of Leaks Encountered - And Their Remedies

1. **Liquid leak.** One basic and very important rule when dealing with chlorine leaks is *always to keep the leak in the vapor phase.* This is usually the case for 150-lb cylinders, since they are stored and usually used in an upright position. With one-ton cylinders, however, liquid chlorine can easily leak through a valve or a fusible plug. Liquid chlorine will vaporize into a gas with a volume approximately 450 times the liquid's.

The leak will be greatly reduced by rolling the ton container (if possible) into a position where gas is escaping instead of liquid chlorine. As the gas escapes, the liquid will refrigerate itself, lowering its vapor pressure and reducing the leakage rate.

2. **Leak at valve packing.** This can be caused by dried-out packings. If so, chlorine will come out from around the valve system and cannot be stopped by tightening down the packing-gland nut. The condition should exist only when valve is opened.

If the leak is very slight, the cylinder should be hooked up and the system should be set to draw chlorine at maximum rate. This should quickly reduce the pressure and probably stop the leak. If leak is of major proportions and does not respond to this treatment then

- the valve should be shut off
- the cylinder should be set outdoors in the shade
- the supplier should be called to pick up the cylinder or replace the packing.

3. **Leak at fusible plug.** This is usually caused by corrosion from moisture, either internally or from the outside.

a. 150-lb cylinders. Leak repair kits are available with a special clamp, a rubber pad, and steel backing. These easily control fusible-plug type leaks. The adapter clamp can also be used to stop this type of leak. A small patch of synthetic rubber gasket material with a follow-up piece of metal are clamped firmly in place.

Note: Putting a clamp over a fusible plug on a 150-lb cylinder leaves it without protection from high temperatures. Use up the chlorine in the cylinder as rapidly as possible.

b. One-ton container (tonner). Leaks can be at the fusible metal or in the thread around plug. If leak is through fusible metal, there is a special clamp of rubber and steel that readily controls the leak. If leak is around the threading of the fusible plug, the Chlorine Inst. emergency kit may have to be used. Since there are six fusible plugs on a tonner, sealing off one of them still leaves five op-

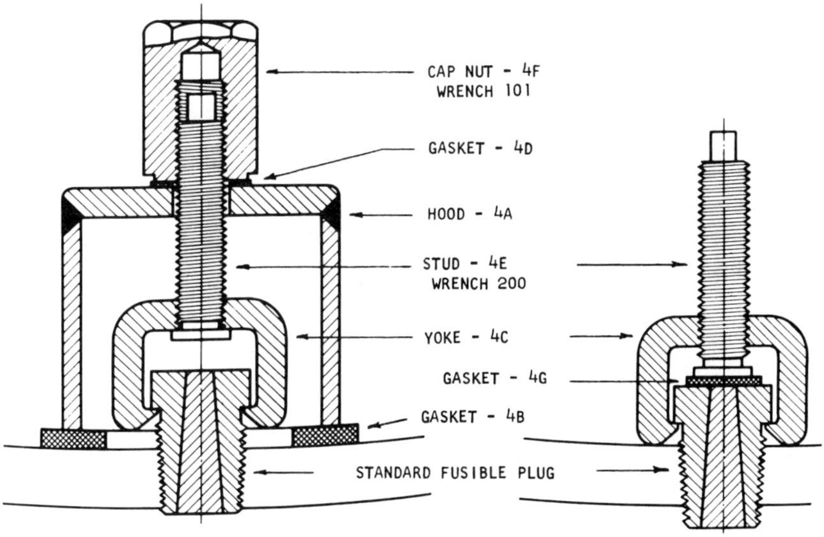

Fig. XIV-5. Device for Sealing Fusible Plug

erating safety devices, which should be enough under most circumstances.

4. **Leaking valve.** One very important word of caution - *anyone taking off the valve outlet cap (on cylinders or tonners) should do it very slowly.* If there is a leak of chlorine past the valve, it will be very noticeable after the cap has been slackened only a single turn. (Very small concentrations of chlorine can be detected by smell.) A mask should be immediately available if required.

5. **As a last resort - contact manufacturer!** If the previously mentioned remedies do not solve the leak problem, the operator should contact the manufacturer. In some cases, a telephone call may provide the operator with the necessary information to correct the leak.

Emergency Plan

If there is a leak from a cylinder that cannot be stopped or repaired, such as a leaking fusible plug, damaged cylinder valve, or a hole in the cylinder body, the following steps should be taken.

1. The operator should protect himself at all times during the emergency, being sure not to be overcome by the leaking gas. (An air pack should be used when in the area of the leaking cylinder.)
2. The operator should contact his supervisor immediately and advise him of problem.
3. If he cannot contact his supervisor, the operator should telephone the fire or police department to tell personnel there of the problem. They in turn should have someone with an auxiliary air pack go to the plant or area of the leak and stand by while the operator tries to stop the leak.
4. If the leak continues, the operator shouldn't allow unchlorinated water into the distribution system. It may be necessary to shut down and use only the water in the reservoir(s). If this is inadequate and the plant cannot be shut down, then the health department should be advised so that health-

department officials can warn residents to boil the water from the tap before use.

5. All events that occur during the emergency should be recorded in the operator's daily log book.

First Aid

It is possible that despite all precautions an employee will suffer contact with chlorine. *Immediate action to counter chlorine's effects is important.*

For first-aid instructions read material contained in first aid section in Chapter XIII.

Appendix
Computing Chlorine Injection

Injection of chlorine is calculated by parts per million (ppm)—that is, so many pounds of chlorine for every million pounds of water. Chlorinator adjustment is usually by pounds per day of chlorine and is calculated on the basis of how many million pounds of water are being treated in 24 hrs and adjusting 1 lb of chlorine for each million pounds of water.

An example is a flow of 10 mgd, which needs to be treated with 2.5 ppm chlorine. One gal of water weighs 8.34 lbs, which, multiplied by 10 million, equals 83.4 million lb. One ppm is 83.4 lbs, 2.5 ppm is 2.5 times 83.4 or 208.5 lbs chlorine for 24 hrs. (It is simpler to solve for 1.0 and then multiply by the desired dosage rate.)

What would be the setting to chlorinate 10 cfs with 1.75 ppm? One cu ft of water weighs about 62.4 lbs. 10 x 62.4 x 3600 (seconds in an hour) x 24 (hours in a day) = 53.9 mil lbs. One ppm would be 53.9 lbs. 1.75 ppm would be 1.75 x 53.9 lb or 94.3 lb chlorine. A shortcut calculation is possible. For 1.0 ppm one can multiply 8.34 times the quantity in million gallons per day. Alternatively the quantity in cubic feet per second can be multiplied by 5.39 (usually this is rounded off to 5.4). In addition, multiplying each 100 gpm (gallons per minute) by 1.2 will also produce the weight of added substance for 1.0 ppm.

Disinfection of a tank or pipeline also involves calculating the weight of water in millions of pounds, then multiplying the number of million pounds of water by the desired dosage. Unless the gallonage is a known figure, it is simpler to use the volume of the tank or section of pipeline in cubic feet and multiply cubic feet by 5.4 for 1.0 ppm. If 50.0 ppm chlorine were needed, one would then multiply by 50. If this disinfection is to be done with 70 per cent available chlorine HTH, obviously more material would be needed to inject the same amount of chlorine: in this case one would divide by 0.70 (70 per cent). Were 0.70 used as a multiplier, the total amount of HTH which has only 70 per cent available chlorine, would be less than if injecting 100 per cent chlorine, which is clearly incorrect.

Chapter XV
Maps, Drawings and Records

Introduction

Maps and records are a vital part of any water distribution system. Without them no utility can operate efficiently and the utility is dependent upon the memories of long-time employees. With each retirement much valuable information can be lost.

Mapping

Most utilities have a large map, called the comprehensive map, that outlines the entire water-distribution system (Table XV-1). This map usually indicates location of water intakes, well fields, reservoirs, trunk lines, fire hydrants, valves and other important structures. Normal practice is to update the map once or twice a year.

Comprehensive maps should always be referenced to the state's plan coordinate system. This enables the accurate location of all points and enables the water utility also to make use of state and federal maps and information that is referenced to the state coordinates.

TABLE XV-1

Items Shown on Comprehensive Map

1. Street names	5. Wells
2. Sizes of mains	6. Orientation arrow
3. Fire hydrants	7. Scale
4. Valves	8. Date last corrected

Section maps, or plats, contain the sort of information shown in Table XV-2. These maps are larger reproductions of sections of the comprehensive map. They are frequently corrected and brought up to date. The scale of these maps may vary from 200 ft/in. for rural areas to 50 ft/in. for congested areas.

Section maps should be numbered and referenced so that the proper map is readily available in case of an emergency. These maps usually also contain references to valve and intersection maps (descriptions of which will follow).

TABLE XV-2
Items Shown on Sectional Plats

1. Plat designation or number
2. Adjacent plat numbers
3. Street names and widths
4. Mains and sizes
5. Material of main
6. Years mains were installed
7. Work orders of main installations
8. Distances from property lines
9. Fire hydrants, numbers, and classifications
10. Valves and numbers (omit valve sheet designation)
11. Valve sheet designation shown in margin
12. Intersection numbers (if valve intersection plats are used)
13. Block numbers
14. Lot numbers
15. House numbers
16. Water account numbers
17. Measurements to service lines
18. Sizes of taps
19. Sizes and materials of service lines
20. Distances, main to stop box
21. Distances, stop box to property line
22. Distances to angle points
23. Distances to fittings
24. Dead ends and measurements
25. Date last corrected
26. Orientation of north arrow
27. Scale

Valve and hydrant maps (Fig. XV-1) show the distance from permanent reference points to each valve and hydrant so that they may be readily located. These maps are carried by persons in the field. These maps provide not only the valve or hydrant locations but also such information as the direction to open, the number of turns to open, the model type, and installation date. Valve and hydrant maps may cover the same area as up to four sectional maps.

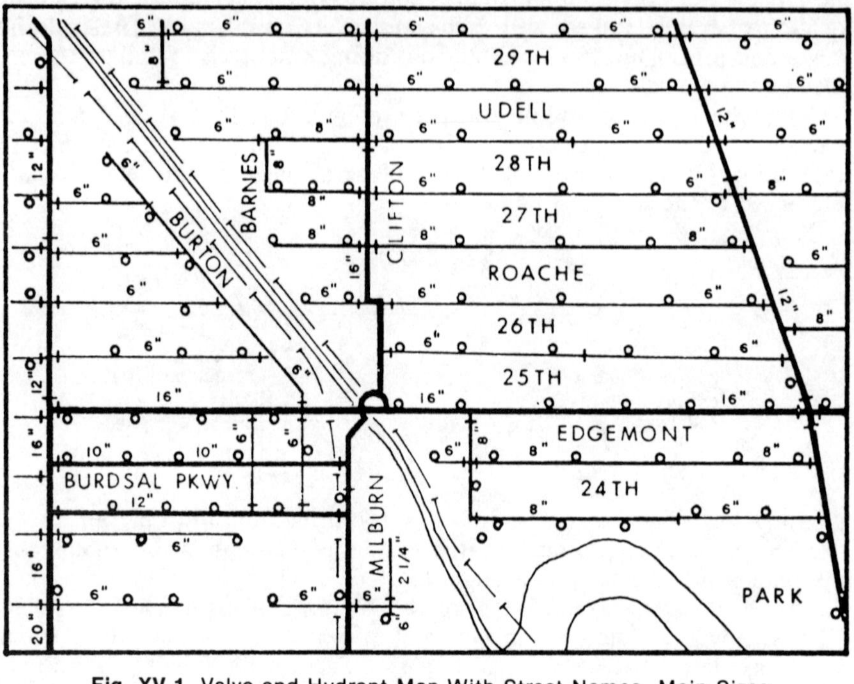

Fig. XV-1. Valve and Hydrant Map With Street Names, Main Sizes

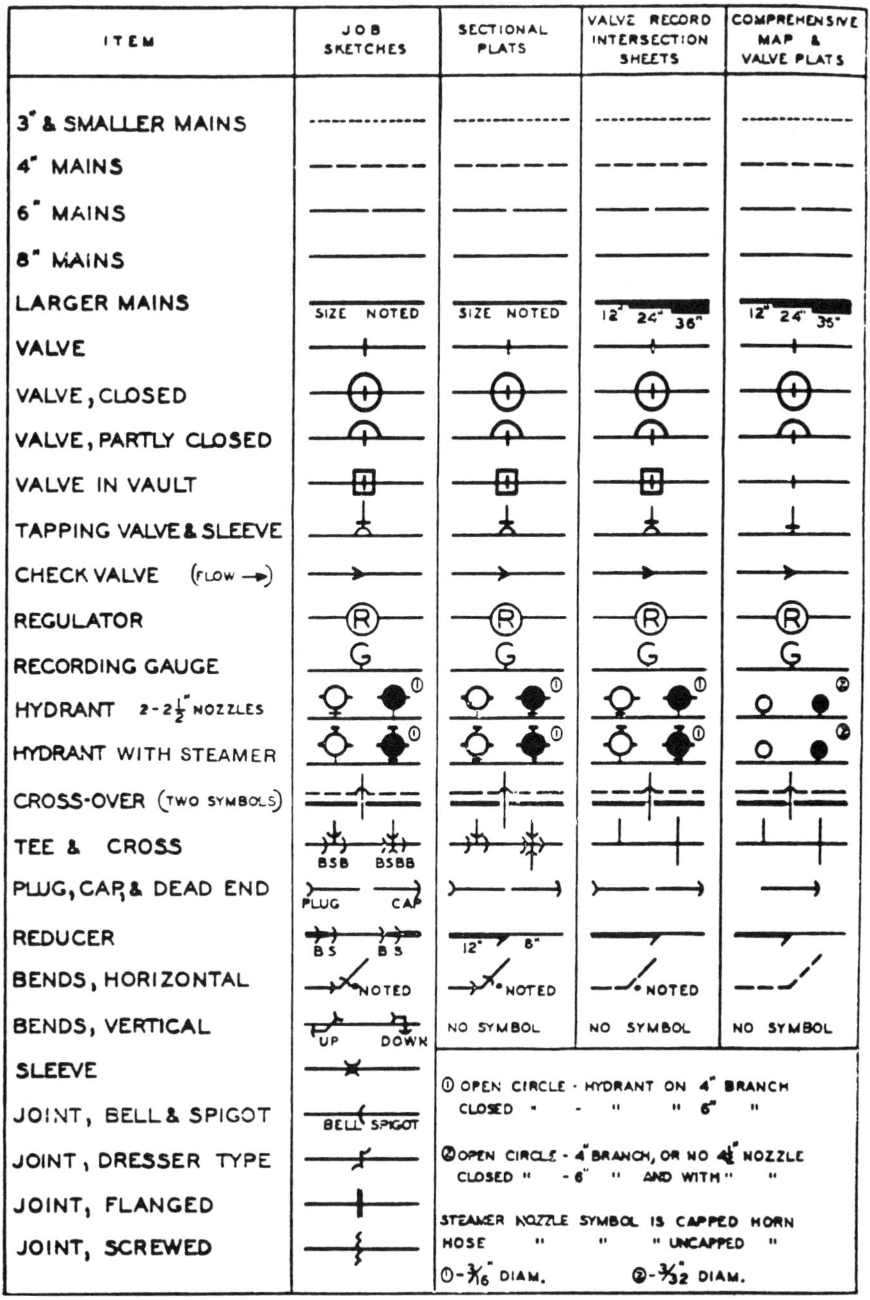

Fig. XV-2. Common Drawing Symbols

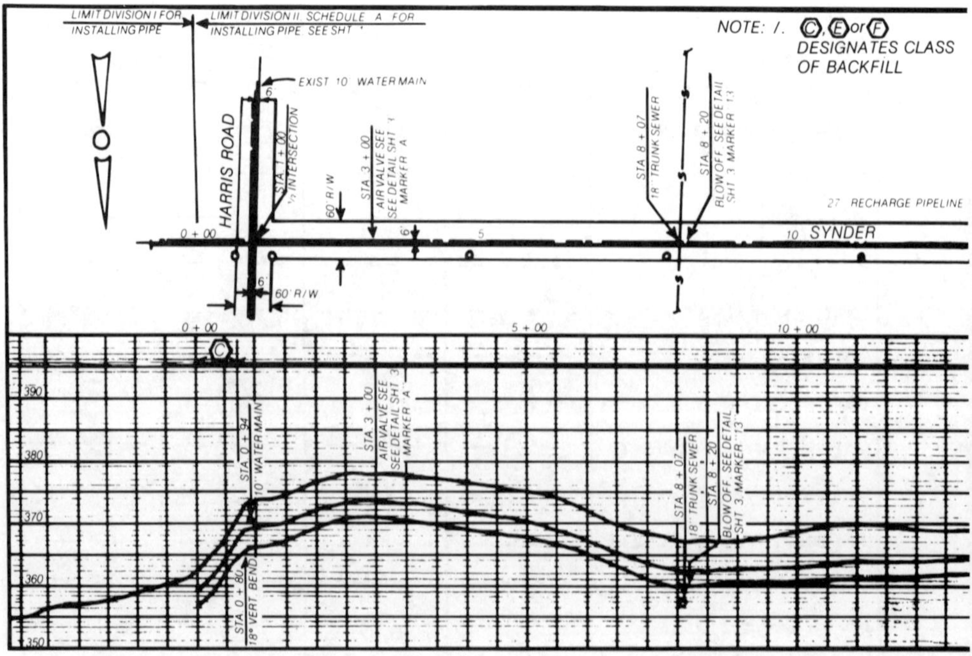
LIMIT DIVISION I FOR INSTALLING PIPE
LIMIT DIVISION II SCHEDULE A FOR INSTALLING PIPE SEE SHT
NOTE: 1. C, E or F DESIGNATES CLASS OF BACKFILL
EXIST 10" WATER MAIN
HARRIS ROAD
STA 1 + 00 INTERSECTION
60' R/W
STA 3 + 00 AIR VALVE SEE SEE DETAIL SHT MARKER A
STA 8 + 07 18" TRUNK SEWER
STA 8 + 20 BLOW OFF SEE DETAIL SHT 3 MARKER 13
27" RECHARGE PIPELINE
SYNDER
0 + 00
5 + 00
10 + 00
390
380
370
360
350
STA 0 + 94 10" WATER MAIN
STA 3 + 00 AIR VALVE SEE SEE DETAIL SHT 3 MARKER "A"
STA 8 + 07 18" TRUNK SEWER
STA 8 + 20 BLOW OFF SEE DETAIL SHT 3 MARKER "13"
STA 0 + 80 18° VERT BEND

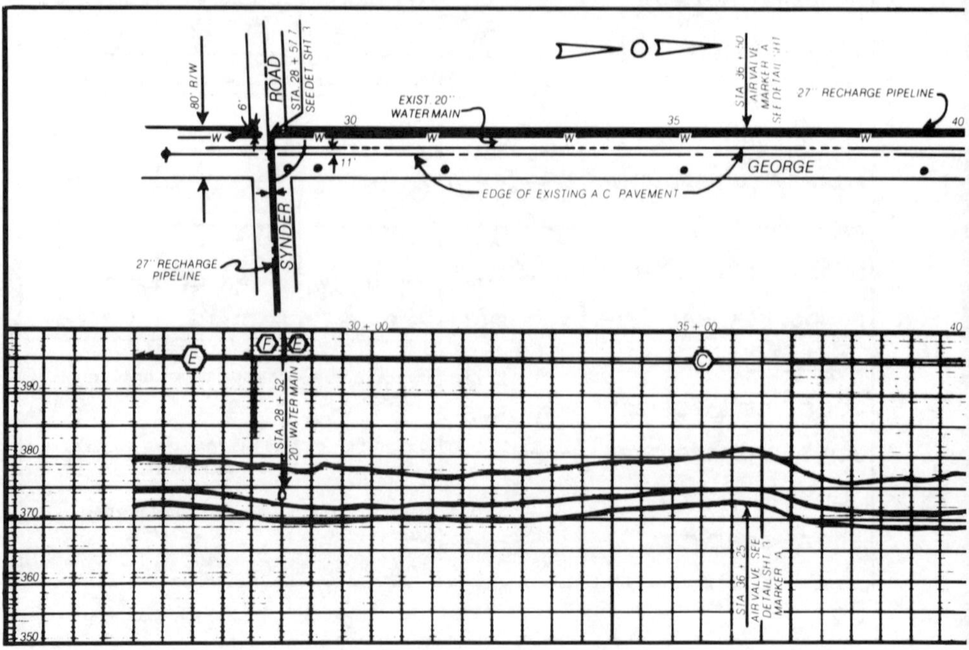
80' R/W
ROAD
STA 28 + 57 SEE DET SHT 3
EXIST 20" WATER MAIN
STA 36 + 40 AIR VALVE MARKER A SEE DETAIL SHT
27" RECHARGE PIPELINE
30
35
40
GEORGE
EDGE OF EXISTING A C PAVEMENT
SYNDER
27" RECHARGE PIPELINE
30 + 00
35 + 00
390
380
370
360
350
STA 28 + 52 20" WATER MAIN
STA 36 + 25 AIR VALVE SEE DETAIL SHT 3 MARKER A

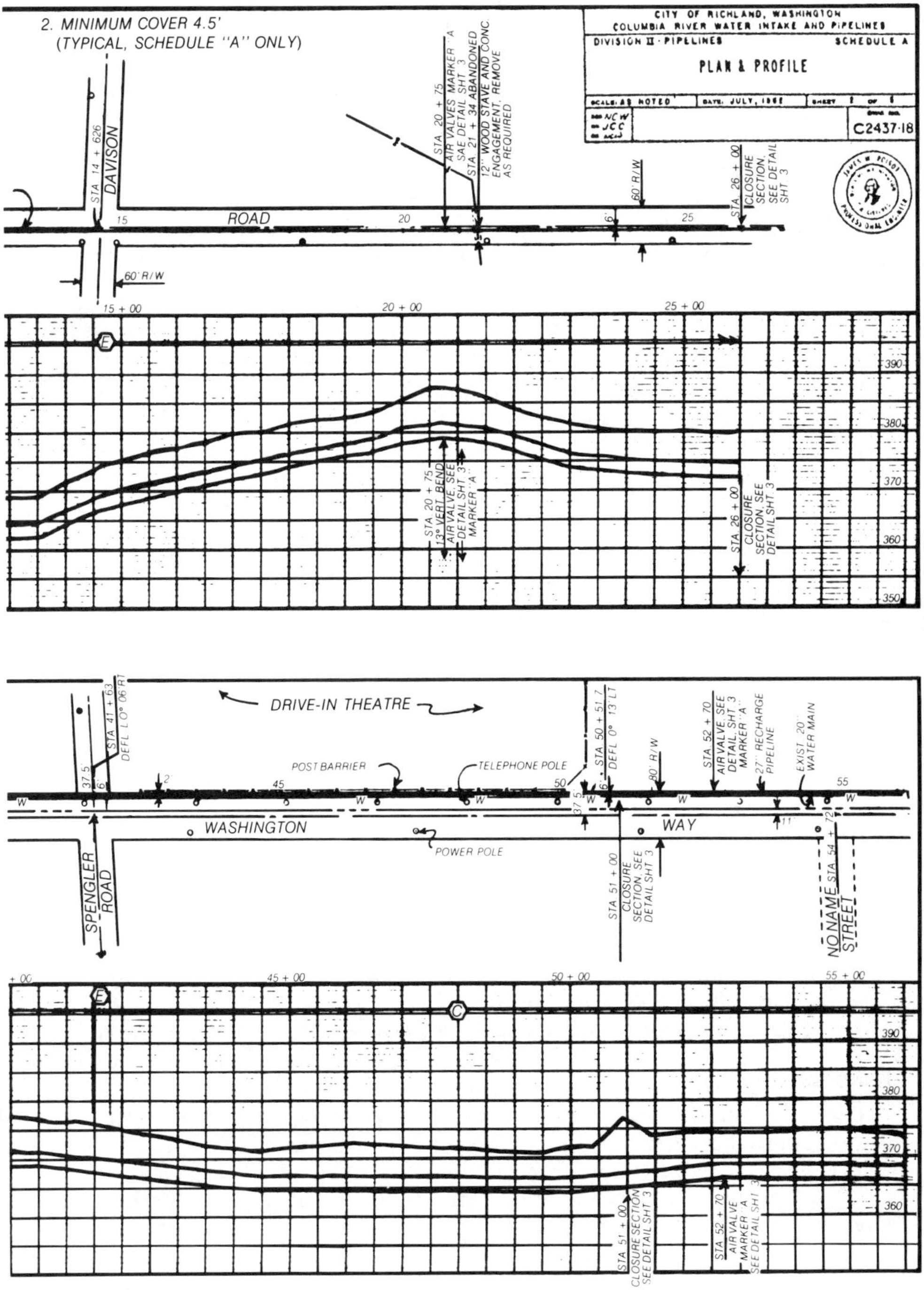

Fig. XV-3. Plan and Profile Maps

Valve and hydrant maps are generally plat and list types. The plat is the map section and shows street names, mains, the size of the main and its designated number, valves, and fire hydrants. Next to the plat is a list that shows valve information such as name, number, size, direction to open, number of turns to open, date installed, references to street curb lines, intersections, property lines, and other remarks.

Many companies use intersection maps rather than valve and hydrant maps. They are referenced by intersection number. However, they are not generally considered as desirable as valve and hydrant maps. Leak-survey maps are frequently used by some districts, but these are usually only modifications of sectional or valve maps.

Every water utility should have a set of standard symbols to denote different items on their maps. A sample of common symbols recommended by AWWA is shown in Fig. XV-2.

Occasionally water operators will need to use plan and profile maps such as in Fig. XV-3. These maps are engineering drawings. Normally they show the depth of the pipe, pipe location, the layout of both horizontal and vertical displacements (curves) and the correct distance from a reference starting point (stationing). Stationing is normally in an engineering form such as 704 + 00, this being the equivalent of 70 400 ft from the starting point.

When line lengths are changed, the affected stations must be given double station numbers so that the system beyond the change does not have to be "restationed." (See Fig. XV-4.) Double station numbers at a location indicate more than one measured distance from the starting point depending on whether the original or changed station is being considered. The two station numbers are listed as being equal to each other to indicate that they are equivalent designations for the same locations.

Common abbreviations found on plan and profile maps include

POT	Point on tangent
POC	Point on curve
BC or (PC)	Beginning of curve
EC or (PT)	End of curve
PI	Point of intersection
EL	Elevation

As-Built Information

During a construction project it is often necessary to change items or design. These changes should not be lost or forgotten. A system should exist to permit the flow of information from the field to the drafting office. This encourages the existence of accurate maps and records of all items in the distribution system.

Records

The most common records in any water agency are those that record information about the water mains, valves, hydrants, and meters.

Most agencies keep an accurate record of the amount of *pipe* laid each month, the type of pipe, and the location of the pipe. Pipelines that are retired from service are also recorded. All repairs and maintenance work to the pipe should also be referenced. All of the information about the mains allows the water utility to estimate present and future needs for new lines or major repairs to old lines.

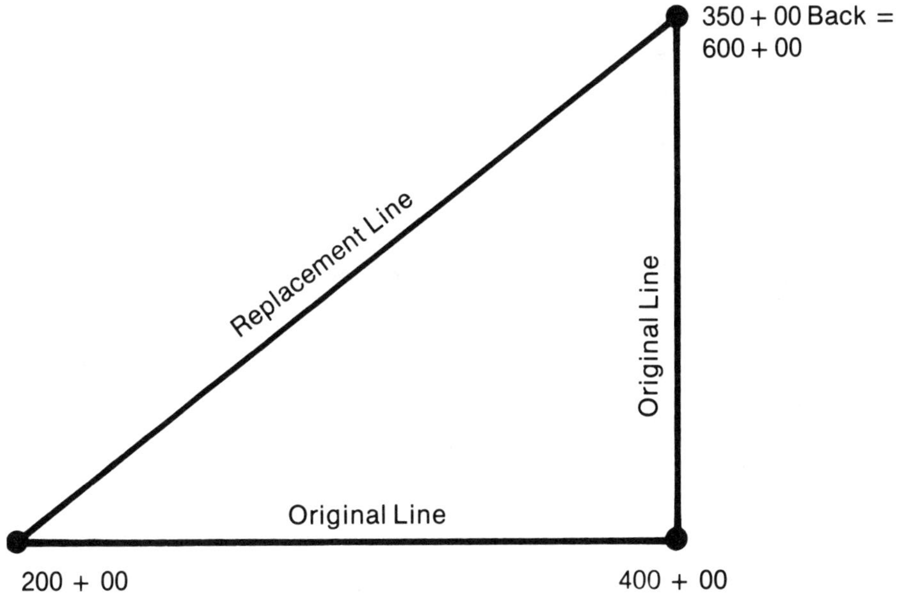

Fig. XV-4. Double-Stationing Scheme

Accurate records should be kept of each *valve* installed in the system, its make, and where and when installed. (See Fig. XV-5.) Many utilities also record each time the valve is moved, and the position of the valve—i.e., open or closed as show in Fig. XV-6.

The location of all *fire hydrants* should be recorded (Fig. XV-5). Also, a complete record should be kept of any work done on a hydrant (Fig. XV-6). If tests are run on the hydrants when in service, then a record should be kept of the test results.

A record should be kept of all *services and meters installed.* Most utilities keep a separate record for each service and for each meter since the meters may be moved around. A record of the type of material, how the connection to the main was made, and the location of all stops should be kept for every service connection. For meters, a record should be kept of all repair work, where installed, the size of the meter, the serial number, and when installed.

Most utilities keep a record of both *quantity of water* produced and the *quality of the water.* The keeping of records regarding the quantity of water produced permits the utility to determine the amount of unaccounted-for water, to estimate future needs for water, and to determine fair water rates. Quality records are vital for the utility. These records can point to ongoing and possible water-quality deterioration, thereby permitting the utility to take preventive action.

Technical Information

Every water utility is at times swamped with technical information that ac-

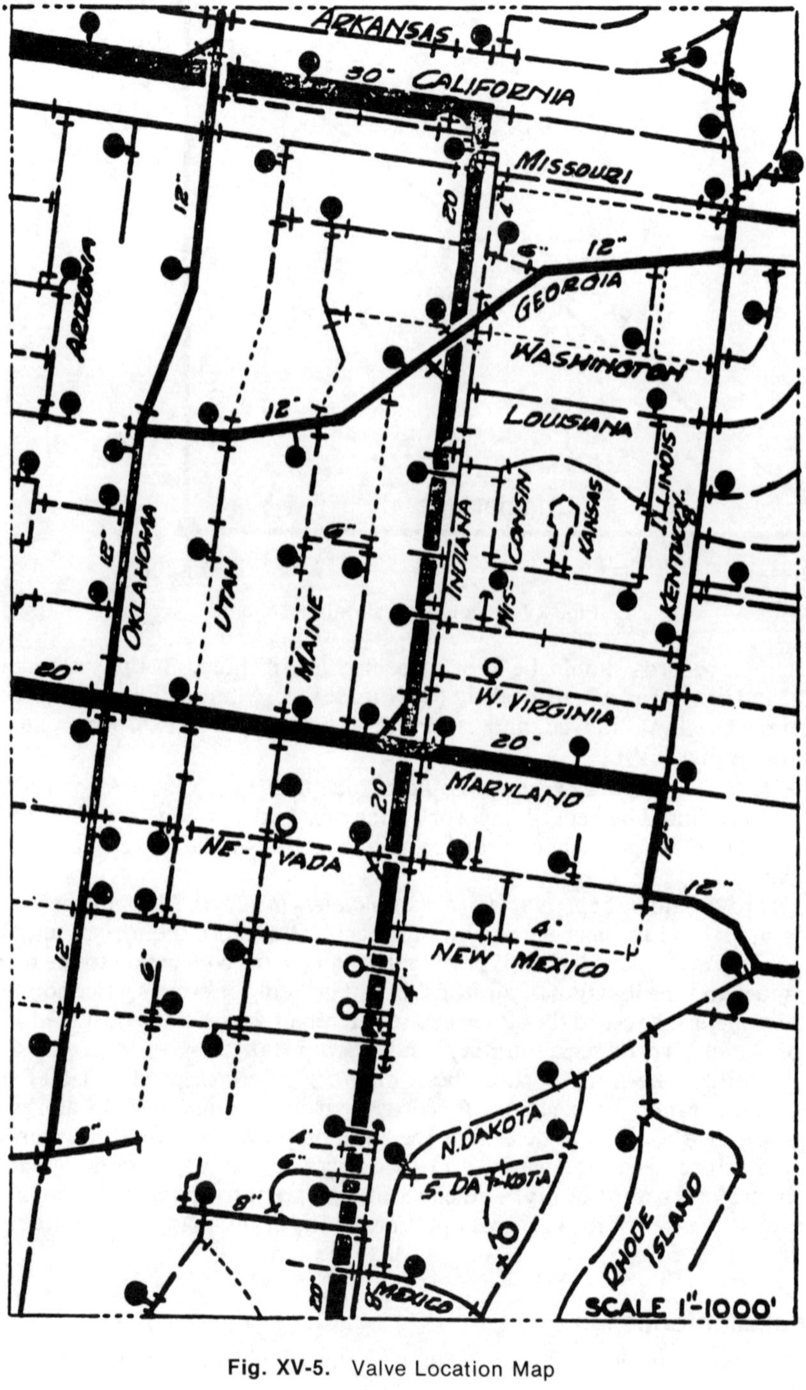

Fig. XV-5. Valve Location Map

ROUTINE VALVE INSPECTION REPORT

(8½" x 11" SHEET)

ROUTINE VALVE INSPECTION REPORT

VALVE PLAT	VALVE NO	1 ST. OPERATION DIR	TURNS	OPENED OR CLOSED	2 ND. OPERATION DIR	TURNS	OPENED OR CLOSED	PACKING	BOX	MEASUREMENTS	REMARKS	OFFICE CHECK
24-1	23	R	22	closed	L	23	opened	OK	OK	OK.		✓
"	24	R	16	closed	L	29	opened	OK	closed 8"	OK	Valve was partly closed.	✓
"	25	R	29	closed	L	23	opened	O.K	OK	Bad.	New measurements taken	✓
"	26	R	20	closed	L	30	opened	O.K.	O.K.	OK		✓
"	27	L	20	opened	R	20	closed	O.K	O.K	O.K.	Valve left closed	✓
"	28	R	50+	?				OK.	OK.	OK.	Stem found broken.	✓
"	29	R.	13	closed	L	13	opened	O.K	O.K	O.K.		✓
"	30	R.	15	closed	L	15½	opened	Tightened	OK.	OK		✓
"	31	R	12	closed	L	12	opened	OK.	OK.	OK.		✓RZ
24-2	17	L	19	closed	R	19½	opened	OK	Lid Missing	Bad.	Lid Replaced New measurements taken	✓RZ
"	18	R	23	closed	L	23	opened	Bad	O.K.	OK.	works hard - stem bent should be replaced	✓RZ
"	27	R	27	opened	Left this valve open			O.K.	O.K.	O.K.	[illegible] valve opens wrong way Change valve record.	✓RZ

XYZ WATER WORKS DATE 9-20-89 BY Jim Jones

VALVE AND HYDRANT INSPECTION REPORT FORMS

VALVE INSPECTION REPORT

MUNICIPALITY ____ VALVE NO. ____

LOCATION ____

SIZE ____ MAKE ____ OPER NUT ____

OPENS ____ NO TURNS ____

VALVE IN ____ ROUTE NO. ____

CONDITION	INSPECTION DATES					
OPENS NO TURNS STEM PACKING NUT BOX or MANHOLE VALVE is NOW INSPECTED BY						

CONDITION	INSPECTION DATES					
OPENS NO TURNS STEM PACKING NUT BOX or MANHOLE VALVE is NOW INSPECTED BY						

CONDITION	INSPECTION DATES					
OPENS NO TURNS STEM PACKING NUT BOX or MANHOLE VALVE is NOW INSPECTED BY						

HYDRANT INSPECTION REPORT

MUNICIPALITY ____ HYD. NO. ____

LOCATION ____

MAKE ____ SIZE ____

2½" NOZZLES ____ STEAMER NOZZLES ____

OPENS ____ SIZE MAIN ____ SIZE BRANCH ____ BRANCH VALVE ____

CONDITION	INSPECTION DATES					
PAINT CAPS CHAINS STEM PACKING DRIP TOP NUT VALVE VALVE SEAT CONDITION of WATER MINUTES FLUSHED STATIC PRESSURE RESIDUAL PRESSURE FLOW INSPECTED BY						

CONDITION	INSPECTION DATES					
PAINT CAPS CHAINS STEM PACKING DRIP TOP NUT VALVE VALVE SEAT CONDITION of WATER MINUTES FLUSHED STATIC PRESSURE RESIDUAL PRESSURE FLOW INSPECTED BY						

Fig. XV-6. Report Forms

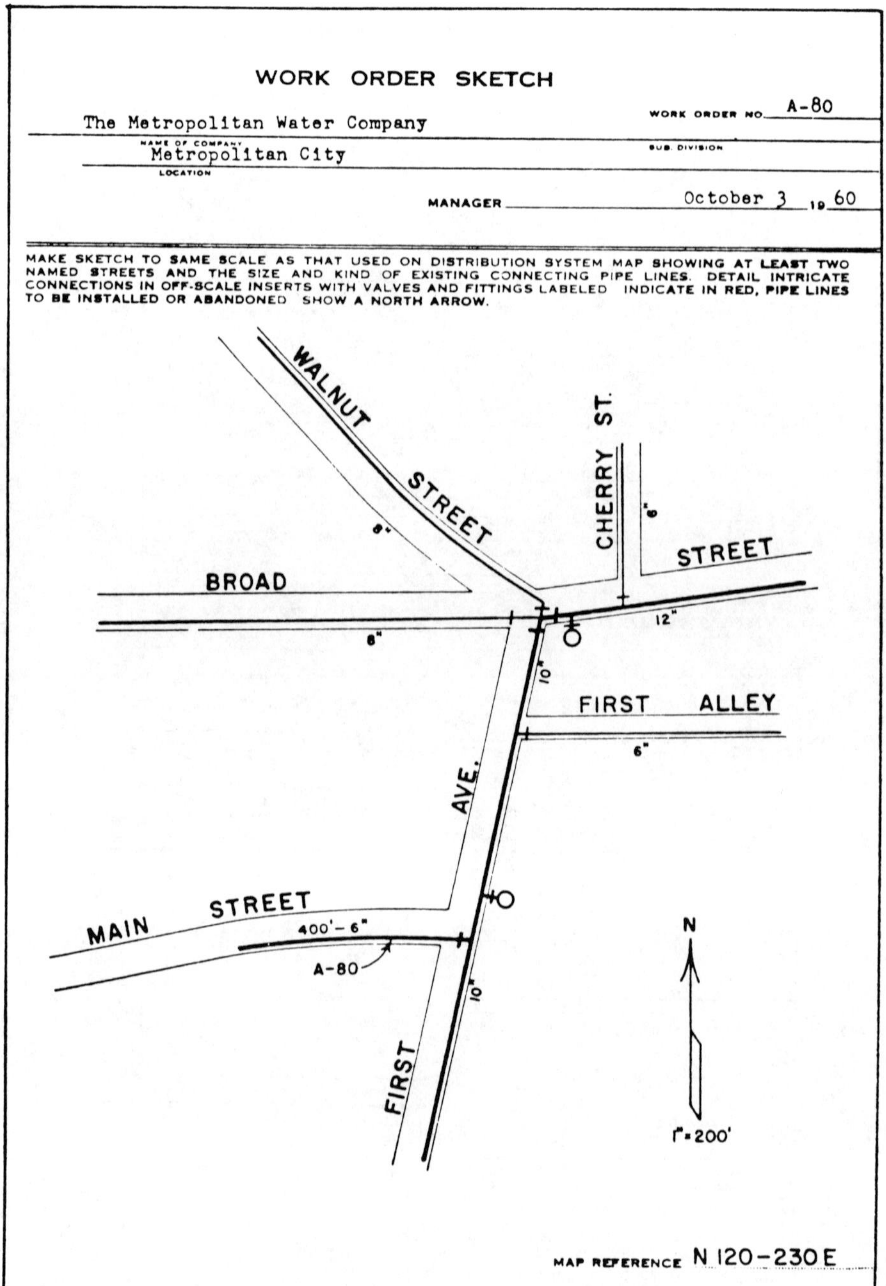
WORK ORDER SKETCH

The Metropolitan Water Company — NAME OF COMPANY

WORK ORDER NO. A-80

Metropolitan City — LOCATION

SUB. DIVISION

MANAGER ______ October 3 19 60

MAKE SKETCH TO SAME SCALE AS THAT USED ON DISTRIBUTION SYSTEM MAP SHOWING AT LEAST TWO NAMED STREETS AND THE SIZE AND KIND OF EXISTING CONNECTING PIPE LINES. DETAIL INTRICATE CONNECTIONS IN OFF-SCALE INSERTS WITH VALVES AND FITTINGS LABELED INDICATE IN RED, PIPE LINES TO BE INSTALLED OR ABANDONED SHOW A NORTH ARROW.

MAP REFERENCE N 120–230 E

Fig. XV-7. Typical Work Order Form

companies the purchase of pumps, meters, and other major items. It is vital to the proper operation of any system and should not be lost or destroyed. In a well-organized system each major item should have a separate file with all of this information recorded, plus any other important documents.

Work Order System

The "work order" system is commonly used in water utilities. This system is designed to ensure that all necessary information is returned to the main office by the construction or maintenance crew. A typical work order form is shown (in Fig. XV-7). These forms permit the keeping of accurate records on the material used and where it was used. "As-built" information may also be recorded on the back of the work order.

Other Records and Maps

Seldom will one find two water utilities using the same record and map systems. However, their bases are almost always the same: the recording of all information that will enable the water utility to operate efficiently.

In the future data processing using computers will find increased use. This should encourage most districts to keep more accurate and up-to-date records since the information will be readily stored, handled, and retrieved.

Persons who are interested in developing a complete set of records for a water utility or in knowing more about them are encouraged to read "Recommended Practice for Distribution System Records" in *Water Distribution Training Course* published by AWWA.

Chapter XVI
Public Relations

What Public Relations Isn't. . .and *Is*

Say "public relations" to most people and they think "publicity" right off the bat. But news publicity is not public relations; it is only a report about the actions of a water supplier and the employees.

Public relations is what the utility's customers think about their water supplier. Their opinion is based on the kinds of services they get from their supplier. And one thing is certain: a utility's public relations never stands still. At any given moment, it is either moving up or down.

Two Keys to Good Public Relations

Because a water utility needs good public relations in order to get support for budgets, bond issues and other policies, it can never take its customers for granted. There are two keys to building good customer opinion. The first is to remember that *everything the utility and its employees do is public relations.* That's right. Everything. The second key is attitude — the *attitude that the customer is always entitled to courteous treatment and an explanation of anything he doesn't understand.*

The meter readers, operators, construction and repair crews, billing clerks, and receptionists — all who work for the water supplier — may never have thought of themselves as public relations representatives, but they are. They are! All of them meet and greet customers at one time or another. And it is those encounters with the public which make the lasting impressions.

Therefore, every contact made with a customer should be considered an opportunity to build good will. Here are some of the ways it can be done.

Employees in the Field — Meter Readers and Repair Crews

The meter reader is the person with whom customers have the most frequent contact. As far as most customers are concerned, the meter reader *is* the utility. And he has one big strike against him the moment he comes to the door: he is associated with the water bill. Thus, he's the utility employee most likely to hear complaints about water rates, service, and system policies. But with careful training and by following sensible rules of conduct, he can become an ambassador of good will.

The meter reader should be neat, courteous. . .and careful! Careful not to track dirt onto clean floors or rugs; careful not to leave the door open; careful not to wake the baby or mistreat the pets; careful not to cause the cake in the oven to fall. Careful, in short, not to cause any undue inconvenience to the customer. And courtesy requires that he properly identifies himself voluntarily.

If the reader suspects a leak on the customer's premises, he can place an immediate check by observing the meter after asking the customer to shut off all fixtures on the premises. If the results warrant it, he can also arrange to have an automatic recording device placed on the meter for a 24-hr period. This provides the customer with a graphic picture of usage during the test period. It can also be used to determine the validity of a high-charge complaint.

By pointing out a leak on the customer's premises, the utility actually does the customer a favor — and makes it clear that it does not want to be paid for water that is not used.

If the utility has a policy of encouraging its customers to read their own meters and send in a report card each month, the meter reader has the opportunity to explain in person how it's done — a special service that can earn real appreciation.

Maintenance and repair crews are the next most "visible" water system people. They have as tough a good-will job as the meter reader does because they are associated with mess, inconvenience, and sometimes danger.

But even these negatives can be turned into good customer relations opportunities.

Like all other utility employees, maintenance personnel should try to remain personally neat on the job. They should practice courtesy in driving and avoid napping in the truck, even if it is on their lunch hour. Napping gives a public impression of the utility being "asleep on the job."

Wherever and whenever possible, the utility should forewarn customers in the area of an interruption in service and, if possible, provide alternate water routing during the repair period. If alternate routing isn't possible, the shutoff should be scheduled during a time of low water demand.

At the job site, courtesy and *safety* are especially important. The repairmen should provide construction signs and barriers. They should keep the area as clean as possible, both during and after the job. And they should repair streets, sidewalks, lawns, yards and fences to look at least as good as they did before the work was begun. Media publicity cannot cover up slipshod performance.

Employees in the Plant — the Operators

The plant operator is the employee whose responsibility is making water safe to drink and use, protecting customer health and welfare. Yet, the customer may hardly know he exists!

Even though the plant operator may have little in-plant contact with customers, he still has his part of play in building good public relations. He should be neat and courteous on the job, particularly if his plant is open for public tours.

In small treatment and distribution systems, the operator is likely to be meter reader, repair crew, and even office manager — all rolled into one person. If so, he then has much contact with his customers and can practice good customer relations as he performs in each job.

Additional Opportunities for Good Company-Customer Relations

Although meter readers and repairmen are most often in the utility's public relations spotlight, other employees have almost unlimited opportunities to build customer good will. Some examples

- The office clerk who is neat, courteous, and helpful to new customers opening accounts
- The employee who handles a customer complaint skillfully, who turns complaints into good-will pluses by being sympathetic, calm, and genuinely interested in the customer's problem
- The employee who provides prompt (preferably immediate!) *action* on a complaint — and then *follows up* to make sure it has been solved
- The custodian who takes special pride in maintaining the utility's plant and reservoir facilities so that they are attractive for customers coming to the offices, touring the plant, or just driving by the reservoir

Even billing can be made a public-relations opportunity. Foremost is to make sure the bill is accurate, itemized, neat, and legible. It should be mailed at the same time every month so the customer can prepare for it and budget accordingly.

The utility can also include a public-relations message with the bill. This can vary from a printed slogan or notice on the bill to separate bulletins or brochures, called "stuffers," included with the bill in the envelope.

Stuffers can contain such information as lists of useful telephone numbers or departments to call for utility services; explanations of company policies; history of the utility; tips on water conservation or how to spot a leak; instructions on how to read the meter; and so on.

Even delinquent billing notices can be tactfully managed. They should always be mailed in envelopes. Their messages need not be irksome. Messages such as "Did our July billing fail to arrive?" or "Did you send payment that we didn't receive?" raise the issue of a missed payment without blaming anyone — not the customer and not the utility. Everyone "saves face" and good customer relations can be maintained.

Employee Relations

The employee who is getting fair treatment in wages, hours, benefits and working conditions will be a public-relations asset whether or not his job ever brings him into direct contact with a customer. He should be kept informed of the utility's purposes, plans and practices, because when he is off the job, *every view he expresses about his utility creates an impression about it in the mind of the listener.*

It's important, therefore, for the utility to be a good place to work — a work environment that *creates in its employees a sense of pride and professionalism.*

Customer Care Summary

In general, "getting along" with customers means

1. Doing routine service and maintenance jobs so well that under normal conditions a customer's water service does not get out of order and if something does go wrong, fixing it quickly and efficiently

2. Accepting routine requests enthusiastically — not as though they are routine, but as though they are as important to the utility as they are to the customer

3. Listening to special requests willingly and looking for a way to say "yes" instead of "no"

4. Listening to complaints sympathetically: if the utility is at fault, making amends; if the customer is wrong or misunderstands the situation, handling the problem so courteously that the customer leaves feeling satisfied, not chastised

5. Making appointments for installations or repair when the customer can conveniently be at home, not simply when workers have time to drop by

6. In general, dealing with customers just as though there were another water company right across the street giving service that is just as good at a price that is just as attractive

At all times the water supplier and its employees should understand what they're doing. . .take pride in their work and appearance. . .be courteous. . . show respect for others and their property. . .and do their share to provide the highest quality water and service for their customers.

Chapter XVII
System Operation

Introduction

In the introduction to this manual it was stated that the water distribution operator is a vital person in providing "Better Water for People." His performance of the tasks assigned is critical for the safe and continuous operation of the system. As he performs he should always be alert to potential problems so that they may be dealt with before they adversely affect the system.

Continuous Positive Pressure

By now the operator should recognize the importance of continuous positive pressure, but he may not be aware of how to affect this variable or how to recognize potential trouble.

An adequate supply of water and a properly designed system can greatly affect pressure performance in a system. Although the distribution operator seldom designs or constructs the water system, he can often minimize system deficiencies by properly operating the system and by making management aware of where critical problems exist. For example, often one line can be extended to another line permitting greater flow of water into an area. But first management must know that there is a problem.

Every system should have a standard location for all sectionalizing valves. These should be readily locatable by all operators, and the valve should be regularly worked and inspected. Valves may have to be uncovered after a street is repaved. By the proper operation of these valves every distribution operator should be able to minimize any out of service areas due to pipeline breaks.

Water Supply

The water supply is a crucial element of the distribution system. The operator should know both where the system derives its supply of water and also how to protect that supply. Customers frequently want to know where their water is coming from. Also, when the system has more than one supply, taste complaints may increase or decrease depending upon the supply used. If the distribution operator can explain these technicalities to the customer, the customer may be reassured about the performance of the water utility.

Where the utility controls its own supply of water, the operator must always be alert to possible contamination of that supply. Activities that must be watch-ed include acts such as damage to the watershed by fire, improper recreational use, improperly installed sewer systems and unauthorized well drilling.

Public Health

We have already stated that the distribution operator is legally responsible for delivering a safe and potable water. The operator can ensure that this occurs by constantly being on the alert for possible events that could contaminate the water. Cross-connection and backflow situations should be apparent to all operators. Unfortunately however, even water districts have been caught with cross connections.

Sampling is an important aspect of every system's operation. The sample must be representative of the water in the system. Whenever a bacteriological problem area is encountered in a system, samples should be taken on a regular and repeated basis until the area returns to normal. Continuous problems in the same area should be thoroughly investigated until a solution to the cause of the problem is reached.

Meters

Every operator should know how to read a meter correctly. Not to be able to do so is to lack a basic skill. Remember that after the meter is read, the customer may also read it. The operator had better be right.

Whenever a new or repaired meter replaces an old or defective meter the customer almost assuredly will want to know why the water bill has gone up. The operator must be prepared to explain to the customer that meters underregister with wear and time. It may be necessary to replace the meter with yet another one to convince the customer. It can save time and money if the meter reader explains that the customer has been getting a reduced charge due to the condition of the meter.

The meter reader should always be on the watch for an underregistering or stopped meter. Checking previous consumption rates is usually the best clue.

Often a customer may complain about lack of water pressure. This is normally due to deficiencies in the customer's own plumbing. The distribution operator should take a pressure reading at the house itself, first without flow to read main pressure, then with flow to show pressure drop due to a plugged line. Often the plumber's valve at the point where the line enters the house has been turned off or is restricting. It is sometimes possible to help the customer by replacing the meter with a larger meter. Where this fails, the owner must usually correct the condition of the household plumbing. Often this only involves replacing the service line from the meter to the house, but repairs to the house plumbing may also be necessary.

In this day and age of expensive plumbers, the operator can be certain that he will be asked questions normally asked of plumbers. The operator should always try to be as helpful as possible but should also be sure of his facts, being careful not to give erroneous or misleading information.

Pump Stations

The distribution operator is almost always responsible for the operation of the pumping stations. The proper functioning of these stations ensures that adequate pressure is always available. Where these stations are located in public areas they should be maintained in such a way that they won't be an eyesore. This normally translates as yearly painting, landscaping, and regular cleaning of the premises. All pumping stations should be fenced to prevent vandalism.

As a distribution operator one should know how to recognize a properly functioning pump and motor. Operations and maintenance procedures are contained in Chapter XI. Standby motors should be started occasionally to ensure that they are operational.

Reservoirs

Every system needs adequate and properly operated reservoirs. A reservoir with low reserves can be a dangerous situation should a fire occur.

Reservoir levels will fluctuate daily. The operator should be able to determine the rate at which the level is changing and know how high the reservoir should be in the morning. This information should permit the operator to tell if the inflow into the reservoir is adequate to meet the system's demand. Input quantities are normally set at the average daily demand to permit the reservoir to float on the system. The operator therefore must determine the average daily demand.

Where the reservoir level is controlled automatically the operator must be able to test the system to ensure that it will properly work when needed. Reservoirs should also have advance warning alarms to signal when levels are too low or too high. The operator should also test them to ensure that they will work when needed.

The operator should always try to spend a few minutes each week visually inspecting the outside of the reservoir. Wet or green areas may be signs of developing leaks. Problem areas detected early usually are not very difficult to repair, but problem areas left unattended may result in the reservoir's being taken out of service.

The grounds around reservoirs should be landscaped and kept clean. This is usually an additional duty for the operator.

Reservoir compounds should be kept locked at all times. This discourages trespassing and tampering with the reservoir. Reservoir hatches or entries should also be kept locked to prevent any sort of intrusion or contamination.

Piping and Easements

The water-distribution operator is normally responsible for patrolling all company pipe lines and easements. During this routine, the operator should watch for unapproved acts within the utility's easement—such as unapproved constuction by other utilities, building by nearby residents, and the crossing of a water line by a sewer line. Also attention should be given to possible damage to facilities by flooding, washing, and fire. Any damage should be noted and reported. Frequent patrolling is a necessity if the utility is not to be taken by surprise.

Security

The operator should certify that the water system does not invite vandalism. Lighting, fences, locks, posting, and frequent patrols all help. Access to the water in a distribution system should be very limited and strictly controlled.

The systems operator has a responsibility for the security of the distribution system, but not to the point of endangering his own life. As difficult as it might be to accept, one should stand by and allow persons to damage facilities or try to disrupt the flow or quality of a water supply as this is the proper course of action according to experts. If an operator's appearance on the scene of vandalism in progress does not discourage the perpetrators, there is not much one can do to be effective in stopping such action.

Normal vandalism usually has little effect on the water system other than the expense for repairs. Persons with intent to disable a water system are usually ineffective since they don't understand the hydraulics involved and the processes and normal safeguards built into most systems, which include chlorine residuals, alternate supplies, system storage, the natural dilution by the large volumes of water involved, parallel-looped grid systems, automatic monitoring and controls, plus materials and techniques available for the treatment of water.

If one discovers vandalism acts in progress, he should make his presence known, but never try to protect the facilities. The operator should immediately notify authorities and observe what has taken place so that he can make an accurate report of the activities. The operator should respond using his knowledge of the operation of the system to isolate the problem. If necessary he should shut off a reservoir if the quality of the water in it is questionable; increase chlorine dosage in questionable water or shut down the system, remembering that a shut down system can pose a serious health hazard in itself. He should then notify customers who might be inconvenienced. Afterward he should prepare to restore affected parts of the water system to normal operation. The operator is a key man in the system operation, especially in an emergency. His well-being is more important than any damage that the system might incur because of acts of vandalism.

If one suspects someone has been tampering with the water or has caused other damage, he should notify responsible authorities, then seek evidence of what might have happened, looking for forced entries, empty containers, missing items, damaged equipment, and any indication of the presence of unauthorized persons, checking with neighbors.

If one fully understands the system, he will know how to operate with part of it disabled. People should know how to contact the operator to report possible damage. The operator should leave word at the agency office where he can be reached in the case of an emergency. Above all he should have an emergency plan, discuss what might happen, correct weaknesses, and if possible, avoid being unprepared.

The Customer

The operator should always perform in a manner that will be appreciated by the customer. He should remember that the customer may not understand operational terms to describe a problem and therefore should remember to explain in terms that the customer will understand.

The customer usually has a twofold relationship with the water utility. First, the customer purchases water from the utility. Second, the customer is a voter who approves or disapproves bond measures in an election. Therefore, even if the customers can't go elsewhere for water, they always have a chance to vote approval or disapproval of the utility.

Training

The role of the distribution operator is so broad that there is always an area in which the operator could use some additional training. Those operators who are constantly improving their performance will always be appreciated both by the company and the public.

System Operation by Crises

In a well-operated system the distribution operator anticipates problems and deals with them before they become a crisis. In a poorly operated system, personnel seem to move from one crisis to another. Preventive maintenance and careful observation, if practiced, can prevent many problems from occuring.

Conclusion

The operator should remember always that he is constantly striving for "Better Water for People."

Glossary

Ammonia - A chemical combination of hydrogen (H) and nitrogen (N) occurring extensively in nature. The combination used in water and wastewater engineering is expressed as NH_3.
Ampere - The unit of flow rate of an electric current.
Anode - Positive pole of an electrode system.
Aquifer - A porous, water-bearing geologic formation. Generally restricted to regions capable of yielding an appreciable supply of water.
Atmospheric Moisture - Water as it occurs in various forms (gas, ice, rain) in the atmosphere.
Axial-Flow Pump - A pump in which the fluid flow is maintained along the direction of the pump axis.
Backfill - (1) The operation of refilling an excavation, usually after some structure has been placed therein. (2) The material placed in an excavation in the process of backfilling.
Backflow - (1) A flow condition, induced by a differential in pressure, that causes the flow of water or other liquid into the distribution system of a potable water supply, from any source or sources other than its intended source. (2) The backing up of water through a conduit or channel in the direction opposite to normal flow.
Bacteria - A group of universally distributed, rigid, essentially unicellular microscopic organisms lacking chlorophyll. Bacteria usually appear as spheroid, rod-like, or curved entities, but occasionally appear as sheets, chains, or branched filaments. Bacteria are usually regarded as plants.
Balanced Valve - A valve that has both sides of the closing mechanism in contact with water and which opens and closes as a result of the difference in water pressures, or water pressures plus other pressures, exerted on the sides of the closing mechanism.
Bituminous - A mixture of hydrocarbons (such as tar).
Booster Station - A pumping station in a water distribution system which is used to increase the pressure in the mains on the discharge side of the main pumps.
Bromine - An element, normally a deep-red liquid, which fumes readily and mixes easily with water. It has been used as a water disinfectant.
Casing - Metal pipe used to seal off a well shaft from the earth surrounding it.
Casing Head - In well boring, a heavy mass of iron screwed into the top of a string of casing to take the blows produced by driving the pipe home.
Cathode - The negative pole of an electrode system which attracts positively charged particles or ions.

Cathodic Protection - An electrical system for prevention of corrosion and pitting of steel and iron surfaces in contact with water, whereby a low-voltage current is made to flow through a liquid or a soil in contact with the metal in such a manner that the external electromotive force renders the metal structure cathodic and transfers the corrosion to auxiliary anodic parts used for the purpose.

Cavitation - (1) The action resulting from forcing a flowing stream to change direction, in which reduced internal pressure causes dissolved gases to expand, creating negative pressure. Cavitation frequently causes pitting of the hydraulic structure affected. (2) The formation of a cavity between the downstream surface of a moving body, for example, the blade of a propeller, and a liquid normally in contact with it.

Centrifugal Force - The apparent force that tends to impel a weight outward from a center of rotation.

Chloramine - A compound of organic or inorganic nitrogen and chlorine.

Chlorinator - A type of equipment utilized for chlorination of water supplies and specifically designed to feed continuously and accurately the desired amount of chlorine or chlorine compound.

Chlorine - An element ordinarily existing as a greenish-yellow gas about 2.5 times as heavy as air. At atmospheric pressure and a temperature of -30F, the gas becomes an amber liquid about 1.5 times as heavy as water. The chemical symbol for chlorine is Cl.

Chlorine Demand - The difference between the amount of chlorine added to water and the amount of residual present. (See "Chlorine Residual.")

Chlorine Residual - Chlorine remaining in water at the end of a specified contact period. Effectiveness of continuing disinfection action depends upon type and amount of residual present.

Cholera - Any of several diseases of man and domestic animals, which may be waterborne, usually marked by severe gastrointestinal symptoms.

Circle of Influence - The horizontal periphery of the depression, roughly conical in shape, produced in a water table by the extraction of water from a well at a given rate.

Clear Well - A reservoir of the storage of filtered water of sufficient capacity to prevent the necessity of frequent variations in the rate of filtration with variations in demand.

Coagulant - A compound responsible for coagulation; a floc-forming agent.

Coagulation - In water and wastewater treatment, the destabilization and initial aggregation of colloidal and finely divided suspended matter by the addition of a floc-forming chemical or by biological processes.

Coliform-Group Bacteria - A group of bacteria predominantly inhabiting the intestines of man or animal, but also occasionally found elsewhere. (See *E. Coli*)

Cone of Depression - The depression, roughly conical in shape, produced in a water table by the extraction of water from a well. The volume of the cone varies with the rate and duration of withdrawal of water. Also called "cone of influence."

Contamination - Any introduction into water of microorganisms, chemicals, wastes, or wastewater in a concentration that makes the water unfit for its intended use.

Continuous Positive Pressure - The maintaining of pressure in a water system. Positive pressure prevents backflow and backsiphonage.
Corporation Cock - A valve for joining a service pipe to a street water main. It is usually owned and operated by the water utility or department. It cannot be operated from the surface. Also called corporation stop, or ferrule.
Cross Connection - (1) A physical connection through which a supply of potable water could be contaminated or polluted. (2) A connection between a supervised potable water supply and an unsupervised supply of unknown potability.
Cycling - The on-and-off running of electrical motors or other equipment.
Detention Time - The time that a slug of water remains in a given holding basin.
Direct-Connected Pump - A power pump which is directly connected to the power unit which operates it by some form of coupling or clutch. Also called close-coupled pump.
Dissolved - Said of a substance thoroughly mixed into another.
Distillation - A process of evaporation and recondensation used for separating liquids into various fractions according to their boiling points or boiling ranges.
Drawdown - (1) The magnitude of the change in surface elevation of a body of water as a result of the withdrawal of water therefrom. (2) The magnitude of the lowering of the water surface in a well, and of the water table adjacent to the well, resulting from the withdrawal of water from the well by plumbing.
Dry Tap - A connection made to a main that is empty (see "Wet Tap").
Ductile - Malleable; capable of being reshaped.
Dysentery - A disease (sometimes waterborne) characterized by severe diarrhea with passage of mucus and blood.
E. Coli - One of the species of bacteria in the coliform group. Its presence is considered indicative of fresh fecal contamination.
Enteric - Having its normal habitat in the intestinal tract of man or animals.
Esthetics - Those qualities of water that deal more with appearance, taste, and odors. These qualities may have no bearing on whether the water is safe to drink but may affect people's desire to drink the water.
Evaporation - The process by which water becomes a vapor at a temperature below the boiling point. The quantity of water that is evaporated.
Evapotranspiration - Water withdrawn from soil by evaporation and/or plant transpiration.
Fecal - Of, or relating to, excrement.
Fire Flow - The rate of flow, usually expressed in gallons per minute, that can be delivered from a water distribution system at a specified residual pressure for fire fighting. When delivery is to fire-department pumpers, the specified residual pressure is generally 20 psi.
Floc - Small gelatinous masses formed in a liquid by the reaction of a coagulant added thereto, through biochemical processes, or by agglomeration.
Flocculation - In water and wastewater treatment, the agglomeration of colloidal and finely divided suspended matter after coagulation by gentle stirring by either mechanical or hydraulic means.
Free Groundwater - Groundwater in the interconnected interstices in the saturation zone; it extends down to the first impervious barrier and moves under the influence of gravity in the direction of the slope of the water table.
Freezing - (1) Converting to a solid state by lowering temperature. Water freezes

at 32F, or 0C. (2) In desalting operations, separation of salt from water by the formation of salt-free ice crystals.

Frost Line - The depth to which the soil will freeze in an area.

Gastroenteritis - Inflamation of the lining membrane of the stomach and the intestines.

Geological Protection - The protection of an aquifer's water quality by impervious strata above the aquifer.

Grid - A network of spaced, perpendicularly set water lines forming a checkerboard pattern.

Ground Water - Water existing beneath the earth in confined or unconfined aquifers.

Hardness - A characteristic of water imparted by salts of calcium, magnesium and iron such as bicarbonates, carbonates, and sulfates, that causes curdling of soap and increased consumption of soap, deposition of scale in boilers, damage in some industrial processes, and sometimes objectionable taste. It may be determined by a standard laboratory procedure or computed from the amounts of calcium and magnesium as well as iron, aluminum, maganese, barium, strontium, and zinc, and is expressed as equivalent calcium carbonate.

Hydraulic Gradient - (1) The slope of the hydraulic grade line; (2) the rate of change of pressure head; (3) the ratio of the loss in the sum of the pressure head and position head to the flow distance. For open channels, it is the slope of the water's surface and is frequently considered parallel to the invert. For closed conduits under pressure, it is the slope of the line joining the elevations to which water would rise in pipes freely vented and under atmospheric pressure.

Hydrologic Cycle - The circuit of water movement from the atmosphere to the earth and returned to the atmosphere through various stages or processes such as precipitation, interception, runoff, infiltration, percolation, storage, evaporation, and transpiration. Also called "water cycle."

Hydropneumatic System - A water system, usually small, wherein a water pump is controlled by the air pressure in a compressed-air tank.

Hygienic - Being conducive to health (i.e. cleanliness).

Hypochlorites - Calcium, sodium, or lithium hypochlorite, which are used for disinfection.

Impervious - Not allowing, or allowing only with great difficulty, the movement of water; impermeable.

Incrustation - The forming of dense solids as a crust on the inside surface of a pipe as a result of hardness or other characteristics of the water carried.

Infectious Hepatitis - An acute virus inflammation of the liver that may be waterborne, characterized by jaundice, fever, nausea, vomiting, and abdominal discomfort.

Infiltration - (1) The flow or movement of water through the interstices or pores of a soil or other porous medium. (2) The quantity of groundwater that leaks into a pipe through joints, porous walls, or breaks. (3) The entrance of water from the ground into a gallery. (4) The absorption of liquid by the soil, either as it falls as precipitation or from a stream flowing over the surface.

Infiltration Gallery - A sizable gallery with openings in its sides and bottom, extending generally horizontally into a water-bearing formation for the purpose of collecting the water contained therein.

Iodine - A nonmetallic halogen element used for disinfection.
Ion Exchange - (1) A chemical process involving reversible interchange of ions between a liquid and a solid with no radical change in structure of the solid. (2) A chemical process in which ions from two different molecules are exchanged.
Leak Survey - A search for leaks in the system, usually done when percentage of unaccounted for water is high.
Manifold - A pipe fitting with numerous branches to convey fluids between a large pipe and several smaller pipes or to permit choice of diverting flow from one of several sources or to one of several discharge points.
Meter Yoke - A fitting especially designed to assist in easy meter installation.
Mixed-Flow Pump - A centrifugal pump in which the head is developed partly by centrifugal force and partly by the lift of the vanes on the liquid. This type of pump has a single inlet impeller; the flow enters axially and leaves axially and radially.
Multi Stage Pump - A centrifugal pump with two or more sets of vanes or impellers connected in series in the same casing.
Mutual Interference - Refers to the overlapping of two or more cones of depression. This results in a reduced output for each well.
Nutating - Wobbling.
Organic Wastes - Wastes derived from living organisms.
Oscillation - A periodic movement to and from, or up and down.
Oxidation - The addition of oxygen to a compound. More generally, any reaction that involves the loss of electrons from an atom.
Ozone - Oxygen in molecular form with three atoms of oxygen forming each molecule (O_3).
Pathogenic - Causing or capable of causing disease.
Pathogens - Pathogenic or disease-producing organisms.
Peak Demand - (1) The maximum momentary load placed on a water plant or pumping station. (2) The average maximum load in one hour or less, but may be specified as instantaneous or with some other short time period.
Petcock - A small cock (valve) used to drain a cylinder, fitting, valve, or similar device.
pH - The negative logarithm of the hydrogen-ion concentration. The concentration is the weight of hydrogen ions, in grams, per liter of solution. Neutral water, for example, has a pH value of 7 and a hydrogen-ion concentration of 10^7.
Poliomyelitis - An acute infectious viral disease considered to be waterborne, characterized by fever, motor paralysis, and atrophy of skeletal muscles often with permanent disability and deformity.
Pollution - A condition representing the presence of harmful or objectionable materials in water.
Potable Water - Water that does not contain pollution, contamination, objectionable minerals, or infective agents and is considered satisfactory for domestic consumption.
Precipitation - (1) The total measurable supply of water received directly from clouds as rain, snow, hail, or sleet; usually expressed as depth in a day, month or year, and designated as daily, monthly, or annual precipitation. (2) The process by which atmospheric moisture is discharged onto a land or water surface.

Pressure Head - The head represented by the expression of pressure over weight (p/y) where p is pressure and y is weight. When p is in pounds per square foot and y is the weight of the liquid per cubic foot, h is head in feet.
Protozoa - Small one-celled animals including amoebae, ciliates, and flagellates.
Pulsating - Functioning in a nonuniform manner.
Pump Efficiency - The ratio of energy converted into useful work to the energy applied to the pump shaft, or the energy difference in the water at the discharge and suction nozzles divided by power input at the pump shaft.
Pumping Level — The elevation at which water stands in a well when the well is being pumped at a given rate.
Quick Operating Valve - A valve that has a revolving plug and a lever attached to it to facilitate sudden opening and closing.
Reaeration — The absorption of oxygen into water under conditions of the water's oxygen deficit.
Recharge — Addition of water to the zone of saturation from precipitation or infiltration from surface streams and other sources.
Register — A device that counts, records or indicates physical quantities or shows variations in magnitude.
Resilient - Capable of withstanding shock without permanent deformation or rupture.
Rotary Drilling — A method of drilling wells in which the drill bit is rotated in the hole. The rock is cut or abraded by knives or hard material set in the bottom of the bit and the waste material is carried away by water or mud forced down the inside of the drill pipe and on up the outside.
Runoff - (1) That portion of the earth's available water supply that is transmitted through natural surface channels. (2) Total quantity of runoff water during a specified time. (3) In the general sense, that portion of the precipitation which is not absorbed by the deep strata but finds its way into the streams after meeting the persistent demands of evapotranspiration.
Safe Yield - The maximum dependable draft that can be made continuously on a source of water supply (surface or ground water) during a period of years during which the probably driest period or period of greatest deficiency in water supply is likely to occur. Dependability is relative and is a function of storage provided and drought probability.
Sampling - The act or technique of selecting a representative part of the water supply for testing.
Sand Trap - A device, often a simple enlargement in cross-sectional area, placed in a conduit to arrest by deposition the sand or silt carried by the water. It usually includes means for ejecting the settled material.
Sanitary Survey - (1) An investigation of any condition that may affect public health. (2) A study of the effect of wastewater discharges on sources of water supply, on bathing or other recreational waters, on shellfish culture and other related environments.
Screen - A device with openings, generally of uniform size, used to retain suspended or floating solids in flowing water and to prevent them from entering an intake or passing a given point in a conduit. The screening element may consist of parallel bars, rods, wires, grating, wire mesh, or perforated plate.

Sectionalizing Valve - A large valve installed in a pipeline to shut off flow in a portion for the purpose of inspection or repair. Such valves are usually installed in the main lines.
Sedimentation - The process of subsidence and deposition of suspended matter carried by water, wastewater, or other liquids, by gravity. It is usually accomplished by reducing the velocity of the liquid below the point at which it can transport the suspended material. Also called "settling."
Seepage - (1) The slow movement of water through small cracks, pores, interstices of a material into or out of a body of surface or subsurface water (2) The loss of water by infiltration from a canal, reservoir or other body of water, or from a field. It is generally expressed as flow volume per unit time.
Silt — (1) Soil particles having diameters between 0.004 and 0.062mm. (2) A size classification of sediment transported by water, air or ice. (3) Deposits of waterborne material in a reservoir, on a delta, or on overflowed lands.
Silting — The process of filling up or raising the bed of a body of water through deposition of sediment.
Sodium Arsenite — A salt used in testing for residual chlorine.
Soil Water — Water existing in the aeration zone, immediately below the surface of the ground. Such water may be discharged into the atmosphere in appreciable quantities by plant transpiration or soil evaporation.
Solvent — Liquid used to dissolve a substance.
Standard Methods — Methods for the examination of water and wastewater published jointly by the American Public Health Association, the American Water Works Association and the Water Pollution Control Federation.
Stand Pipe — (1) A pipe or tank connected to a closed conduit and extending to or above the hydraulic grade line of the conduit. It is often installed to afford relief from surges of pressure in pipelines. (2) A tank resting on the ground having height greater than diameter and used for storage of water in distribution systems.
Static Level — (1) The elevation of the water table or pressure surface when it is not influenced by pumping or other forms of extraction from the ground water body. (2) The level of elevation to which the top of a column of water would rise, if afforded opportunity to do so, from an artesian aquifer, or basin, or a conduit under pressure. Also called hydrostatic level.
Storage Effect — The improvement of water quality because of the water's being stored. Normally this is due to the settling out of material and bacteria.
Strata — Sheetlike masses of sedimentary rock or earth of one kind lying between beds of other kinds.
Surges — A momentary increase in flow (in an open conduit) or pressure (in a closed conduit) that passes longitudinally along the conduit, usually due to sudden changes in velocity.
Tapping Sleeves — A split sleeve used in making a wet connection where a single branch line is to be tapped into a water main under pressure.
Toxic — Of, relating to, or caused by a poison.
Transpiration — (1) The process by which water vapor is lost to the atmosphere from living plants (2) The quantity of water thus dissipated.
Tuberculation - The formation of tubercules or nodules in pipe, with a resulting increase in frictional coefficient.

Turbidity - (1) A condition in water or wastewater caused by the presence of suspended matter, resulting in the scattering and absorption of light rays. (2) A measure of fine suspended matter in liquids. (3) An analytical quantity usually reported in arbitrary turbidity units determined by measurements of light diffraction.

Typhoid - A communicable disease that may be waterborne, marked especially by fever, diarrhea, prostration, headache, and intestinal inflammation. Caused by *Salmonella typhosa* bacterium.

Ultraviolet Irradiation - Very short wave lengths of light not visible to the human eye, sometimes used for disinfection.

Valve Boxes - A metal or concrete box or vault set over a valve stem and rising to the ground surface, to allow access to the stem in opening and closing the valve. A cover is usually provided at the surface to keep out dirt and debris.

Vapor - (1) The gaseous form of any substance. (2) A visible exhalation, such as fog, mist, steam, smoke, diffused through, or suspended in, the air.

Vaults - Underground structure, normally made of concrete.

Velocity Head - The vertical distance or height through which water must fall freely, under the force of gravity, to acquire the velocity that it possesses. It is theoretically equal to the square of the velocity divided by twice the acceleration of gravity.

Viruses - The smallest (10-300 m μ m in diameter) form capable of producing infection and diseases in man or other large species. The true viruses are insensitive to antibiotics. They multiply only in living cells. They do not multiply by division as do intra-cellular bacteria.

Volute Pump - A centrifugal pump with a casing made in the form of a spiral or volute as an aid to the partial conversion of the velocity energy into pressure head as the water leaves the impellers.

Water Hammer - The phenomenon of oscillations in the pressure of water about its normal pressure in a closed conduit, flowing full, which results from a too-rapid retardation of flow. Momentary pressures greatly in excess of the normal static pressure may be produced in a closed conduit by this phenomenon.

Water Shed — (1) The area contained within a divide above a specified point on a stream. In water supply engineering, it is called a watershed or catchment area; in river control engineering it is called a drainage area, a drainage basin, or a catchment area. (2) The divide between drainage basins.

Wet Tap — A connection made to a main that is full or under pressure.

Bibliography

A Training Course in Water Distribution—M8. AWWA, Denver (1962).

Action Now Part I, II & III, California Dept. of Health, Sacramento (1970).

Advanced First Aid and Emergency Care. The American National Red Cross, Doubleday & Co., Garden City (1st ed., 1973).

Angele, Gustave. *Cross Connections and Backflow Prevention Handbook.* AWWA, Denver (2nd ed., 1974).

Basic Water Treatment Operator's Manual—M18. AWWA, Denver (1971).

Basic Gas Chlorination Workshop. Ministry of the Envir., Toronto (1974).

Basic Manual Water Plant Operators, Part I. Oregon State Univ. and Oregon State Bd. of Health, Salem (2nd ed., 1968).

Basic Manual Water Plant Operators, Part II. Oregon State Univ. and Oregon State Bd. of Health, Salem (2nd ed., 1968).

Chlorine Manual. Chlorine Inst., New York (4th ed., 1969).

Corrosion Control Practices. AWWA, California Sect. Los Angeles (2nd ed., 1974).

Domestic Water Supply. California Dept. of Health, Sacramento (1974).

Fundamentals of Water Supply Operation. State of Illinois (1972).

Glossary, Water and Wastewater Control Engineering. Published by AWWA, APHA, ASCE, & WPCF, Denver (2nd ed., 1969).

Graham, Kennard. *Fundamentals of Electricity.* American Technical Soc. (5th ed., 1968).

Hirsch, A.A. *Manual for Water Plant Operators.* Chemical Publishing Co., New York (1945).

Improving Water Treatment Operations—S107. AWWA, Denver (1970).

Kirkpatrick, Joanne. *Mathematics for Water and Wastewater Treatment Plant Operators. (Book 1—Fundamental)* Ann Arbor Science Publishers, Inc. Ann Arbor (1973).

Leupold & Stevens. *Stevens Water Resources Data Book.* (1st ed., 1974).

Manual of Instruction for Water Treatment Plant Operators. New York State Dept. of Health, Albany (1975).

Manual of Water Utility Operations. Texas Water Utilities Assn., Austin (6th ed., 1975).

National Interim Primary Drinking Water Regulations. Federal Register FRL 464-7, Title 40, Chapter 1, Subchapter D.

Recommended Practice for—M14 Backflow Prevention and Cross-Connection Control. AWWA, Denver (2nd ed., 1974).

Renn, Charles E. *Investigating Water Problems.* LaMotte Chemical Co. Chestertown (1970).

Riehl, Merrill L. *Water Supply and Treatment.* National Lime Assn. Washington D.C. (11th ed., 1976).

Safety Practice for Water Utilities—M3. AWWA, Denver (1971).

Standard First Aid and Personal Safety. The American National Red Cross, Doubleday & Co., Garden City (1st ed., 1973).

The Story of Water Supply. AWWA, Denver (1969).

Studybook for Wastewater Operator Certification. WPCF, Washington, D.C. (1974).

Suggested Course Outline for Water Distribution System Operators, Grade I. AWWA, Denver (1967).

Suggested Course Outline for Water Distribution System Operators, Grade II. AWWA, Denver (1967).

Suggested Course Outline for Water Distribution System Operators, Grade III. AWWA, Denver (1967).

Suggested Training Outlines for Water Treatment Operators, Grade I. AWWA, Denver (1970).

Supervisors Safety Manual. National Safety Council, Chicago (3rd ed., 1972).

Sweitzer, Robert F., and Flentje, Martin E. *Basic Water Works Management,* American Concrete Pressure Pipe Assn. Washington, D.C. (1972).

Symons, George E. *Pipes and Piping.* Dun Donnelley Publishing Co., New York, (1967).

Symons, George E. *Pumps and Pumping.* Dun Donnelley Publishing Co., New York (1966).

Waste Water Reclamation. California Dept. of Water Resources, Sacramento (1973).

Wastewater Reclamation: Instrumentation & Controls. California Dept. of Health, Sacramento (1970).

Water Chlorination Principles and Practices—M20. AWWA, Denver (1973).

Water Distribution Operator Training Curriculum, Revised Outline Grades I & II. AWWA California Sect., Los Angeles (1975).

Water Meters—Selection, Installation, Testing, and Maintenance—M6. AWWA, Denver (2nd ed., 1973).

Water Well Standards. California Dept. of Water Resources, Sacramento (1968).

Waterworks Manual, Basic. Coordinating Council for Occupational Education, Washington, D.C. (1972).

Water Works Operators Manual. Alabama Dept. of Public Health, Montgomery (3rd ed., 1972).

Notes

Notes

Notes

Notes